Chettapong Janya-Anurak

Framework for Analysis and Identification of Nonlinear Distributed Parameter Systems using Bayesian Uncertainty Quantification based on Generalized Polynomial Chaos

Karlsruher Schriften zur Anthropomatik
Band 31
Herausgeber: Prof. Dr.-Ing. Jürgen Beyerer

Eine Übersicht aller bisher in dieser Schriftenreihe
erschienenen Bände finden Sie am Ende des Buchs.

Framework for Analysis and Identification of Nonlinear Distributed Parameter Systems using Bayesian Uncertainty Quantification based on Generalized Polynomial Chaos

by
Chettapong Janya-Anurak

KIT Scientific Publishing

Dissertation, Karlsruher Institut für Technologie (KIT)
Fakultät für Informatik, 2016

Impressum

KIT Scientific Publishing

Karlsruher Institut für Technologie (KIT)
KIT Scientific Publishing
Straße am Forum 2
D-76131 Karlsruhe

KIT Scientific Publishing is a registered trademark of Karlsruhe Institute of Technology. Reprint using the book cover is not allowed.

www.ksp.kit.edu

Print on Demand 2017 – Gedruckt auf FSC-zertifiziertem Papier

ISSN 1863-6489
ISBN 978-3-7315-0642-3
DOI 10.5445/KSP/1000066940

Framework for Analysis and Identification of Nonlinear Distributed Parameter Systems using Bayesian Uncertainty Quantification based on Generalized Polynomial Chaos

zur Erlangung des akademischen Grades eines

Doktors der Ingenieurwissenschaften

der Fakultät für Informatik
des Karlsruher Instituts für Technologie (KIT)

genehmigte

Dissertation

von

Chettapong Janya-anurak

aus Bangkok, Thailand

Tag der mündlichen Prüfung: 14. 11. 2016

Erster Gutachter: Prof. Dr.-Ing. Jürgen Beyerer

Zweiter Gutachter: Prof. Dr. Hermann G. Matthies

Acknowledgement

The present work was accomplished while being a research assistant at the Lehrstuhl für Interaktive Echtzeitsysteme (IES) at the Karlsruhe Institute of Technology (KIT) in close collaboration with the Mess-, und Regelungs und Diagnosesysteme (MRD) department at Fraunhofer Insitute of Optronics, System Technologies and Image Exploitation (IOSB). Over the past years this work received benefits from several people and it would not have been possible without them. I would like to cordially thank all those who have been directly or indirectly involved with this work.

First and foremost, I would like to express my sincere gratitude to my "Doktorvater" (advisor) Prof. Dr.-Ing. Jürgen Beyerer for his continuous support of my Ph.D study, for encouragement on my research, and for allowing me to succeed and "be born" as one of the "made in Germany" Dr.-Ing. I would like to thank him for always dedicated his time to discuss my idea and issues despite the busy schedule as director of IES and Fraunhofer IOSB. His knowledge, research suggestion and direction which flows in this thesis, as well as his exemplary mentoring as "doctoral father", have been invaluable for me.

I would also like to express my thank to Prof. Dr. Hermann G. Matthies for his interest in my work and for agreeing to act as co-advisor. I would like to thank him for his scientific guidance and the inspiring discussions since our first meeting in Porguerolles, which broadened my research perspectives and provided invaluable knowledge throughout my research. Especially, the conversation before my examination as well as "Hans im Glück", that he fabled, will always be in my memory.

Special thanks are given to the MRD department at Fraunhofer IOSB. Especially my sincere thanks also go to Dr. Thomas Bernard, my group leader, who provided me an opportunity to join his team since my first internship until my doctoral graduation. Without his guidance and his cooperation with the industrial partner, my research would not have come into fruition. I would like to thank Prof.Dr.-Ing. Michael Heizmann and also Christian Frey as the department manager for all organizational and scientific supports.

I would also like to thank all MRD colleagues, IES doctoral fellows and also former colleagues for their great friendship, for supporting me in writing this thesis and encouraging me to strive towards my goal. In particular, Thomas Stephan, Peter Frühberger, Mathias Ziebarth, Sebastian Höfer, Mathias Braun, Masoud Roschani, Christian Frese, Mahsa Mohammadikaji, Philipp Woock and also the secretary of MRD, Birgit Kugler. Special thanks go to Christian Kühnert, my office mate, for daily conversations in our office and for his instruction on, how I should manage my life with a new-born baby.

A special thanks go to all proofreaders of this work. Without the valuable feedback, this work would not be at the quality that it now is.

Graduate studying, especially, abroad can sometimes be a long and solitude process. I am very fortunate and grateful to all Thai friends in Germany and in Karlsruhe for making my living in Germany not as lonely as it could be. Especially, I would like to thank Ekkachai Thawinan, who shares the same fate as me as a Thai Ph.D. student at KIT, for all stimulating discussions and mathematical explanations. I would like to thank Thiti Sirithanakorn for his suggestions, carefully proofreading my thesis and also his acting as host at the weekend before my examination. I would like to thank Alexander Loefflath and his family for his supporting during my stay in Germany, I am very appreciated to meet you again in Germany since our school days.

Through the wonders of the internet, I was able to keep the contact to my friends and my colleagues in Thailand. I would like to thank all of them for their support and encouraging. Especially, I would like to thank my schoolmates, Jak Tanthana, for the support in English language.

Last but not least; I especially would like to thank my family. I deeply thank my parents for their spiritual supporting, encouragement and endless patience. It was their love that underpinned my persistence in the graduate career and picked me up when I got weary. My sister has been very supportive and taking care of my parents throughout my study in Germany. At the end I would like to express appreciation to my beloved wife, Duangporn Janya-anurak. Words cannot express how thankful I am to her for all of the sacrifices that she has made on my behalf. Our wonderful daughter, Alisa, her smiles encourage me to overcome the difficulties encountered during my pursuit of the doctoral degree. Your unfailing love, understanding and support make the completion of this thesis possible.

Karlsruhe, February 2017 Chettapong Janya-anurak

Kurzfassung

Viele Industrie- und Umweltprozesse sind komplexe örtlich-zeitliche Systeme, die auch als verteilt-parametrische Systeme (engl. *distributed parameter system*; DPS) bezeichnet werden. Sie können oft durch nichtlineare gekoppelte partielle Differentialgleichungen (engl. partial differential equation; PDE) beschrieben werden. Solche Systeme sind oft komplex und es ist schwierig die Beziehung zwischen Modelleingang, Modellausgang und den Parametern herzustellen. Zusätzlich weicht die Prädiktion der physikalischen Modelle häufig von den realen Messungen ab. Das Ziel der Arbeit ist es, einen systematischen Ansatz zu finden, um das Systemverhalten zu verstehen und die Abweichung zwischen der Modellprädiktion und dem realen Prozess zu reduzieren.

In den meisten realen Anwendungen besitzen die Modellbenutzer Vorwissen über das System. Dieses Vorwissen kann als Unsicherheit im Bayes'schen Sinne interpretiert werden. Die Unsicherheiten werden durch Zufallsvariablen oder ein Zufallsfeld und mit der Methode der Unsicherheitsquantifizierung (engl. *Uncertainty Quantification*; UQ) beschrieben.

Dazu werden geeignete UQ Ansätze ausgewählt und kombiniert, mit denen das DPS analysiert und identifiziert werden kann. Das Framework verwendet dazu die Unsicherheitspropagierung (engl. Uncertainty propagation; UP), die Sensitivitätsanalyse (SA) und die Bayes- Parameterkalibrierung. Die Hauptschwierigkeit der Anwendung dieser UQ Ansätze auf nichtlineare DPS liegt im Rechenaufwand. Herkömmliche Verfahren, wie beispielweise sampling-basierte Verfahren, benötigen zahlreiche Modellberechnungen. Dieser Rechenaufwand kann durch das verallgemeinerte

Polynomial Chaos (engl. generalized polynomial chaos; gPC) vermindert werden. Die gPC Ansätze sind als effiziente Methode zur Quantifizierung der Unsicherheit bekannt.

Die statistischen Kennwerte und das globale Sensitivitätsmaß, nämlich Sobol Indizes (engl. *Sobol Indices*), können durch die gPC-Koeffizienten unmittelbar berechnet werden. Damit kann die gPC-Approximation als Ersatzmodell verwendet werden, um den Rechenaufwand von umfangreichen Modellen zu reduzieren. In Rahmen dieser Dissertation wird das gPC-Ersatzmodel auf die Unsicherheitspropagierung, die lokale Sensitivitätsanalyse und das Bayes'sche inverse Probleme angewendet. Als ein neues numerisches Verfahren wird die *Polynomial Chaos Expansion* (PCE) auf die rekursive Bayes'sche Schätzung angewendet. Diese Überlegung führt zu einem deterministischen, linearen Filter, welches in dieser Dissertation mit *linear Bayesian updating with PCE* bezeichnet wird.

Die Effizienz des in dieser Arbeit vorgestellten Frameworks wird anhand eines realen industriellen Prozesses gezeigt. Bei diesem industriellen Prozess handelt es sich um ein komplexes rheologisches Verfahren zur Herstellung von Glasröhren und -stäben, welche Vorprodukte (Preforms) für optische Fasern sind. Mit Hilfe des Frameworks kann der Prozess systematisch analysiert und die Modellparameter optimal kalibriert werden, um die vom Anwender definierten Leistungskriterien zu erfüllen.

Abstract

Many industrial and environmental processes are characterized as complex spatio-temporal systems. Such systems known as *distributed parameter systems* (DPSs), which are often modeled with nonlinear coupled partial differential equations, are usually highly complex and it is difficult to establish the relation between model inputs, model output and parameters. Most importantly, the solutions of physics-based models commonly differ from the real measurements. Hence, the aim of the thesis is to elucidate a systematic framework that allows to gain an understanding of the behavior of nonlinear distributed parameter system. Ultimately, the framework reduces the disagreement between the computational model predictions and the measurements of the real processes.

For most of real world applications, the model users have some background knowledge about the considered system. This available background knowledge is exploited in this thesis, by formulating the state of knowledge in form of uncertainties in Bayesian sense. The uncertainties are described by random variables or random field and are quantified by *Uncertainty Quantification* (UQ) approaches to gain the knowledge about the system.

The appropriate UQ approaches are selected and combined systematically in order to analyze and identify systems. The framework proposed in this thesis includes the uncertainty propagation, the sensitivity analysis and the Bayesian parameter calibration. The main challenge of applying the UQ to the nonlinear distributed parameter systems is the computational efforts. By the conventional method such as sampling-based method, it requires numerous simulation evaluations. This computational effort is diminished by the *generalized polynomial chaos expansion* (gPCE), which has been proposed as an efficient methodology for uncertainty quantification.

The statistical values of the system responses and the global sensitivity measures (Sobol Indices) can be computed directly from the gPC coefficients. A surrogate model (with gPC approximation) is used to reduce the computational effort of extensive models. In this thesis, the surrogate model is applied to the uncertainty propagation, to local sensitivity calculation, and to the Bayesian inverse problem. The polynomial chaos expansion (PCE) can also be applied to the recursive Bayesian estimation resulting in an deterministic linear filter, in this thesis named as *linear Bayesian updating with PCE.*

The efficiency of the framework is assessed in the application to a model of a real industrial system. The considered industrial process is a complex rheological forming process producing glass tubes and glass rods, which are pre-products for optical fibers. With this application, the framework is illustrated to be capable of systematically analyzing the system and optimally calibrating the model parameters to fulfill user defined performance criteria.

Table of symbols

General notation

x	Scalar
$\boldsymbol{x}$	Vector
$\mathbf{x}$	Block-vector
$\mathbf{X}$	Block-matrix
$\hat{x}$	Approximation of x
$X = X(\omega)$	Random variable
$\boldsymbol{X} = \boldsymbol{X}(\omega)$	Random vector

Symbols

d	Dimension of random variables
e	Residual, Error

i, j, k	Generic counting index
t	Time
$\boldsymbol{r}$	Spatial coordinate $[x\ y\ z]^T$
$\boldsymbol{v}$	Velocity vector $[v_x\ v_y\ v_z]^T$
$\boldsymbol{n}$	Normal vector
$x(\boldsymbol{r}, t)$	System state
$\mathbf{x}_\infty$	Stationary solution of $\mathbf{x}$
$\boldsymbol{y}$	Measurement
$\boldsymbol{y}^M$	Model prediction
β_k	PCE coefficients
w	Measurement noise
$\boldsymbol{q}$	Parameter vector
$\boldsymbol{Q}(\omega)$	Random vector representing uncertain parameters
$\mathcal{X}(\boldsymbol{r}, \omega)$	Random field
$\mathcal{X}(t, \omega)$	Stochastic process
$\phi(\boldsymbol{r})$	Spatial basis function
$\varphi(\boldsymbol{r})$	Test function
$\psi(\xi)$	Univariate PC basis function
$\Psi(\boldsymbol{\xi})$	Multivariate PC basis function
$\Xi_P(\cdot)$	Polynomial Chaos of order P
$\gamma_k = \|\Psi_k\|^2$	Square norms of the PC basis function Ψ_k
$H(\cdot)$	Hermite PC basis function
$\boldsymbol{H}(\cdot)$	Multi-dimensional Hermite PC basis function
$La(\cdot)$	Lagauerre PC basis function

$Le(\cdot)$	Legendre PC basis function
$L(\cdot)$	Likelihood function
$\mathbf{B}$	Matrix of PCE coefficients
$\mathbf{I}$	Identity matrix
$\mathbf{J}$	Jacobi matrix
$\mathbf{K}$	Kalman gain
N_P	Number of PCE terms
N_s	Number of spatial basis functions
N_t	Number of time increments
N_Q	Number of quadrature nodes
P	Expansion polynomial order
PE	Polynomial exactness
$N_{Q_l^d}$	Number of d-dimensional quadrature points accuracy level l
N_{KLE}	Number of truncated terms of KLE
N_{MC}	Number of samples from Monte-Carlo algorithm
N_{samp}	Number of samples
N_{step}	Number of update steps
N_r	Number of regression points
$\mathcal{P}$	Probability measure
$\mathfrak{A}$	Event space
Ω	Sample space
ω	Event in the sample space Ω
$P_X(x)$	Cumulative distribution function of X
$p_X(x)$	Probability density function of X

$p(q,y)$	Joint probability density function of q and y
$p(q\|y)$	Conditional probability density function of q given y
$\mathbb{E}[X(\omega)]$	Expectation of $X(\omega)$
$\overline{X},\mu_X$	Mean of $X(\omega)$
σ_X	Standard deviation of $X(\omega)$
σ^2	Variance
$\sigma^2_{\mathcal{A}}$	Partial Variance of $\mathcal{A}$
$\mathcal{N}(\mu,\sigma^2)$	Normal distribution with mean μ and variance σ^2
S_i	Sensitivity or Sobol index to a parameter q_i
$\mathbf{C}_{\boldsymbol{XY}}$	Cross covariance matrix between $\boldsymbol{X}$ and $\boldsymbol{Y}$
$\mathbf{C}_{\boldsymbol{X}} = \mathbf{C}_{\boldsymbol{XX}}$	Covariance matrix of $\boldsymbol{X}$
$\mathbf{R}_{\boldsymbol{X}}$	Correlation matrix of $\boldsymbol{X}$
$\mathcal{Q}^d_l f$	d-dimensional quadrature level l of integrand f
$\mathcal{S}(L,d)$	Sparse grid quadrature for level L and dimension d
w^i	Weight of quadrature formulas
$\mathcal{G}$	Spatial domain
$\mathcal{T}$	Time interval of interest
$\partial\mathcal{G}$	Boundary of the spatial domain $\mathcal{G}$
$\mathcal{L}_p$	The space of the Bochner-Lebesgue p-integrable functions
$:=$	defined as
$\langle\cdot\|\cdot\rangle$	Inner product
∇	Nabla operator $\nabla = \frac{\partial}{\partial x}\ \frac{\partial}{\partial y}\ \frac{\partial}{\partial z}$
$\otimes$	Tensor product

Symbols in neutron diffusion example

φ	Neutron flux	in neutrons $\text{cm}^{-2}\,\text{s}^{-1}$
S	Distributed source term	in neutrons $\text{cm}^{-3}\,\text{s}^{-1}$
D	Diffusion coefficient	in cm
Σ_a	Macroscopic absorption cross section	in cm^{-1}
Σ_d	Detector's equivalent reaction cross section	in cm^{-1}

Symbols of the glass forming model

$\boldsymbol{\sigma}$	Cauchy stress tensor	in Nm^{-2}
F_{vis}	Viscous force	in N
$MPpc$	Mass throughput	in %
q''_{rad}	Radiative heat flux	in W $(\text{m})^{-2}$
$t_{\text{ss\_end}}$	Ending time of the stationary phase	in s
$t_{\text{ss\_start}}$	Starting time of the stationary phase	in s
z_{s1}	Sensor position	in m
g	Gravitation constant	in $\text{m}\,\text{s}^{-2}$
σ_B	Stefan-Boltzmann constant	in $\text{W}\,\text{m}^{-2}\,\text{K}^{-4}$
ϵ	Emissivity	in 1
ΔD	Shrinkage	in mm
A	Cross section area	in mm^2
D	Diameter	in mm
D_{cyl}	Diameter of glass cylinder	in mm

R	Radius	in mm
W	Wall thickness	in mm
η	Dynamic viscosity	in Pa s
λ	Effective heat transfer coefficient (conductivity)	in W $(\mathrm{m\,K})^{-1}$
ρ	Density	in kg m^{-3}
c_p	Specific heat capacity at constant pressure	in J $\mathrm{kg}^{-1}\,\mathrm{K}^{-1}$
$\boldsymbol{v}_{\mathrm{feed}}$	Feeding speed	in mm s^{-1}
$\boldsymbol{v}_{\mathrm{pull}}$	Pulling speed	in mm s^{-1}
T	Temperature	in degC
T_g	Glass temperature	in degC
T_{amb}	Ambient temperature	in degC
T_{cyl}	Temperature of glass cylinder	in degC
$T_{\mathrm{oven\_max}}$	Maximal oven temperature	in degC
T_{oven}	Oven temperature	in degC

Glossary

1D	One dimensional
2D	Two dimensional
3D	Three dimensional
AD	Automatic Differentiation
ALE	Arbitrary Lagrangian-Eulerian
BC	Boundary Condition
BEST-EST	Best-Estimate Model Calibration

BVP	Boundary Value Problem
CFD	Computational Fluid Dynamics
DPS	Distributed Parameter System
DRAM	Delay Rejection Adaptive Metropolis-Hasting
FDM	Finite Difference Method
FEM	Finite Element Method
GHQ	Gauss-Hermite Quadrature
GMKF	Gauss-Markov-Kalman Filter
gPC	generalized Polynomial Chaos
gPCE	Generalized Polynomial Chaos Expansion
GSA	Global Sensitivity Analysis
IBVP	Initial Boundary Value Problem
IC	Initial Condition
IVP	Initial Value Problem
KF	Kalman Filter
KLD	Kullback-Leibler Divergence
KLE	Karhunen Loève Expansion
LHS	Latin Hypercube Sampling
LSA	Local Sensitivity Analysis
M&S	Modeling and Simulation
MaxEnt	Maximum Entropy Principle
MC	Monte-Carlo
MCMC	Markov-Chain-Monte-Carlo
MOR	Model Order Reduction

NISP	Non-intrusive spectral projection
ODE	Ordinary Differential Equation
PC	Polynomial Chaos
PCE	Polynomial Chaos Expansion
PDE	Partial Differential Equation
pdf	Probability Density Function
QMC	Quasi-Monte-Carlo
QoI	Quantities of Interest
RV	Random Variable
SA	Sensitivity Analysis
SDE	Stochastic Differential Equation
SSM	Stochastic Spectral Method
UP	Uncertainty Propagation
UQ	Uncertainty Quantification
V&V	Verification and Validation
VFT	Vogel-Fulcher-Tammann
WRM	Weighted Residual Method

Contents

1 Introduction

1.1 Motivation

In the modern era, the scientific computing has played a large role in natural sciences and engineering disciplines. The scientific computing based on mathematical models is an essential tool for engineers and scientists to analyze, design and control technical processes. The mathematical models appear in many forms, such as dynamical systems, statistical models, game theoretic models, etc. The model types are chosen based on the behavior of the processes and the task to be accomplished.

Many industrial and environmental processes have spatio-temporal interrelations, as their process variables vary both temporally and spatially. Such systems are known as spatially distributed parameter systems (DPSs). Modeling of DPS is essential for process control, prediction and analysis [Li11]. Many DPSs, such as thermal process, fluid process and transport-reaction process, can be described mathematically with Partial Differential Equations (PDEs). In practice, modeling a real-world process often leads to nonlinear coupled PDEs. Because analytical solutions often are not readily available for such models, the mathematical model is translated into a numerical model as a computer code for simulations. Modeling and simulation (M&S) are important topics in the advanced scientific computing currently. A computational model can be applied to gain knowledge about the process, which then can be used to improve the analysis, design and control such process.

However, the two following issues are usually encountered when applying the computational model.

1. All models are imperfect abstractions of reality. In real-world process, uncertainties are inherent and arise from various factors. These imperfections that are unaccounted in the modeling leads to an uncertainty in predicted values and in discrepancies between the solution of physics-based models and the results empirically determined from the real processes.

2. Most models, especially those based on nonlinear coupled PDEs considered in this thesis, are often highly complex. The physical manifestations of such systems are difficult, since establishing the relation between inputs, outputs and parameters are not straightforward. The lack of deep understanding of the process also reduces the effectiveness of the model used to design and optimize the control strategies of the process.

The two aforementioned issues have to be taken into account. In order to apply a computer simulation to investigate various aspects of the process efficiently. In practice, these two issues are treated rather based on the experience, training and knowledge of the model user.

Hence, the purpose of this work is to elucidate a systematic approach that allows one to gain an understanding of the behavior of nonlinear distributed parameter system and reducing the discrepancy between the computational model predictions and the measurements acquired from the real process.

1.2 Related work

Since M&S are broadly utilized in natural sciences and engineering disciplines, numerous approaches have been proposed in various scientific communities to handle the two mentioned problems.

Aiming at the knowledge gain as the main objective, the exploitation of M&S to obtain the knowledge about the process can be schematically pictured as shown in figure 1.1.

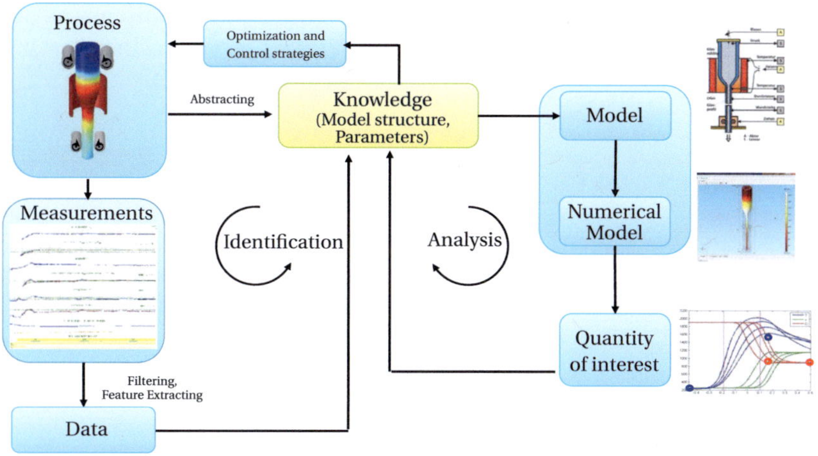

Figure 1.1: Schematic of the application of the modeling and simulation to harvest the knowledge

In general, engineers and scientists have some empirical knowledge about the process, which usually appear in form of mathematical equations, empirical assumptions, or certain ranges of parameter values. These empirical knowledge is used to establish a computational model. The knowledge is normally acquired through operational experience, process observation and repeated measurements. The approaches, which the model user employs to gain the new knowledge and deal with the two mentioned problems are the following: First, **analyzing** the system via computer simulation provides a system behavior insights, which facilitates the understanding of the system. Second, **identification** through observations and measurements can also provide insights of the system.

In the following section, the overview of related topics relevant to the thesis according to these two issues is briefly discussed.

1.2.1 System identification

Verification and Validation in scientific computing

The scope of modeling and simulation (M&S) can be divided into three major components, namely the modeling, the numerical and the measurement data components. The relationship between these three components and the role of the *Verification and Validation* (V&V) is presented in figure 1.2.

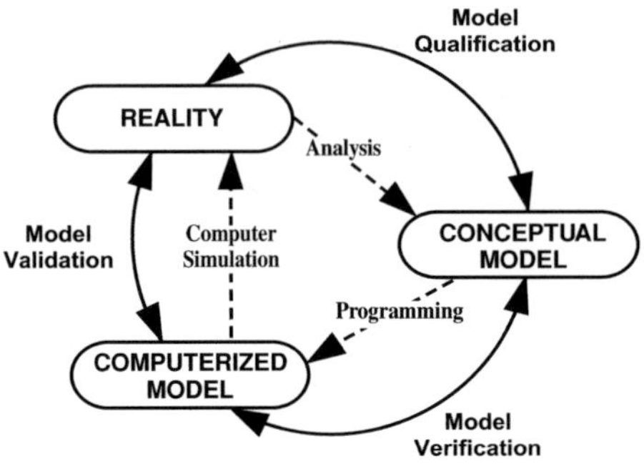

Figure 1.2: Phases of modeling and simulation and the role of verification and validation [Sch79]

Verification and Validation (V&V) are the primary processes for assessing the accuracy and the credibility of computational results. In the context of scientific computer simulation, verification of a model is the process that approve the correctness of implementation based on the conceptual model. While the validation is the process of assessing the credibility of mathematical model based on comparisons between computational results and experimental data. The validation is also used to determine the validity range of the model.

V&V are important approaches to check the agreement between the model predictions and the experimental data. The *Verification and Validation* (V&V) could be considered as the closest idea to the framework proposed in this thesis. The main objective of V&V is rather the approving of the correctness of the implemented model and conceptual model, it does not aim to obtain knowledge about the system. More details about the verification and validation can be found in [Obe98b, Obe98a, Obe10, Roy11, Roa97].

Parameter estimation

The discrepancy between the model prediction and the perfect measurement can be solved by determining the model parameter values, which allow the predicted result from computational model to be close to the real-world measured data. In the scope of mathematical modeling, this procedure is known as the parameter estimation problem or sometimes referred as the parameter calibration problem. The parameter calibration is one of the selected approaches applied in this thesis and therefore is essential topics which is discussed more in detail in the chapter 5.

1.2.2 System analysis

Computer Experiment

One of the simplest ways to study the system can be done by computing the model at some specific deterministic parameters. This procedure, also known as a computer experiment, is widely employed in engineering and natural sciences. A computer experiment serves the model user to analyze the system based on the model responses with various parameters. The set of deterministic parameters can be chosen arbitrarily by the users to investigate the model response according to the change of these parameters. In some cases, the set of deterministic parameters are systematically determined with the optimal designs approach, which is considered to be a class of *design of experiment* (DOE) approaches. The optimal design constructs the set of parameters to provide optimal properties in some senses. The details about design of experiment for computer simulation can be found in [San13] and [Puk93]

Sensitivity Analysis

The problem setting in the experimental design is very similar to that in the field of sensitivity analysis. In both disciplines, one tries to obtain information from the system with a minimum of physical or numerical experiments. The term *sensitivity analysis*, however, has different connotations in various scientific communities. The objective of sensitivity analysis can be viewed as quantifying the "relative" contributions from individual parameters and determining how variations in these parameters affect the system responses. In the course of the computer simulation, which is the scope of this thesis, the sensitivity analysis is also used to gain the information about the intrinsic relationships in the model. Sensitivity analysis is one of the main topics in this thesis, therefore, the state-of-the-art pertain to this topic will be discussed more in detail in the chapter 4.

Simulation under the influence of uncertainties

Another topic that is closely related to sensitivity analysis is the uncertainty propagation in the model. As mentioned before, all models are only imperfect abstractions of reality. Significant uncertainties are inherent in real systems and consist in the model. Uncertainties can arise from many sources. They can be addressed in three groups according to the three components in the modeling and simulation: (a) the incomplete knowledge for constructing the conceptual modeling including the simplification assumptions for mathematical models, (b) the numerical errors in the computational model and (c) the inherent noise in the measurement data. By simulating under the influence of the uncertainties, these uncertainties are propagated as variation in the simulation results.

A sensitivity analysis aims to identify which factors, i.e. variables or parameters, make significant impact on the system response, while the uncertainty propagation tries to describe the entire set of possible outcomes, together with their associated probabilities of occurrence. In many publications, the propagation of uncertainty can be found under other terms, such as uncertainty analysis or uncertainty quantification. There is no precise definition for this terminology. In this thesis, the term *uncertainty propagation* is used in this context and the term *Uncertainty Quantification* (UQ) is defined according to [Smi13] as the science of identifying, quantifying and reducing uncertainties associated with models, numerical algorithms, experiments and predicted quantities of interest.

Quantifying uncertainty is attaching a measure to the uncertainties [Mat07]. This can be accomplished by representing the uncertainties using appropriate mathematical formulations. For example the *evidence theory* [Obe05], the *worst-case scenarios* [DC95, Han92], the method based on fuzzy set theory [Eli00] and the probabilistic theory. Expressing the uncertainties with a probabilistic description seems to be mostly adopted in practice, because it offers most information compared to others [Mat07]. The stochastic approach of uncertainty modeling is achieved by representing uncertainties in the models as random variables, stochastic processes or random fields.

1.3 Contributions

For most real world applications, the model users have some background knowledge about the considered system. Such available background knowledge should be exploited, in order to to improve understanding of the system behavior and reduce the discrepancy between the computational model prediction and the measurement of the real processes.

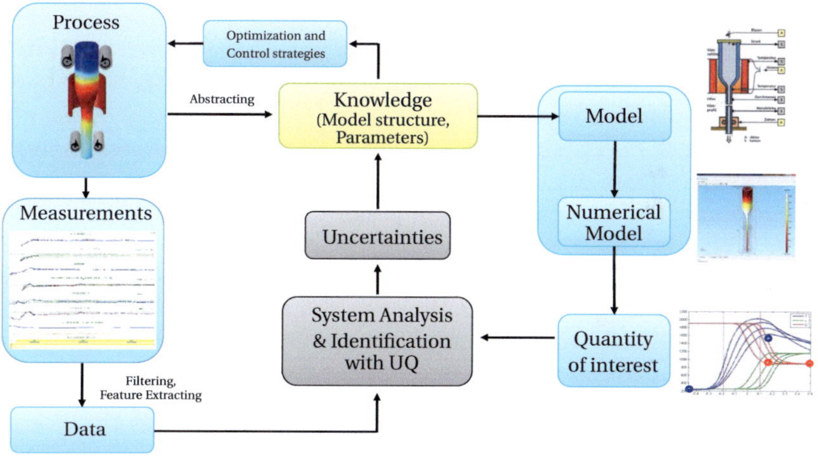

Figure 1.3: Schematic of the framework for analysis and identification using Bayesian uncertainty quantification

The available knowledge can be exploited by using Bayesian formalism to interpret the incomplete knowledge in form of uncertainties. In the Bayesian sense, the probability is used to express the subject's ignorance, the incomplete knowledge can be represented in form of epistemic uncertainties, which can be described by the random variables in the stochastic approach. Then the uncertainties can be quantified by using various probabilistic *Uncertainty Quantification* (UQ) approaches to improve the knowledge. These procedures are combined in the form of a framework as shown in figure 1.3.

This thesis presents the framework for analysis and identification nonlinear DPS using Bayesian UQ approach as a main contribution. The framework is devised for the DPSs. In case of DPSs the uncertainties can be modeled in form of stochastic process or random field. Solving the problem of DPSs required an appropriate discretization methods, which are discussed in this thesis as well.

The appropriate UQ approaches are selected and combined systematically in order to analyze and identify systems. By applying the UQ to the nonlinear DPSs, many issues have to be considered and resolved. The main challenge of applying the UQ to the DPSs is the computational effort. Simulations under influence of uncertainty parameters or solving the inverse problem for parameter calibration often require numerous simulation evaluations. For such cases of computationally intensive models, a single simulation can take many minutes, hours or even days to complete. Many mathematical approaches, e.g. *Model Order Reduction, surrogate model,* have been proposed to reduce the computational effort. In the last decade, the spectral method, such as the *generalized Polynomial Chaos* (gPC), has been proposed as an efficient methodology for UQ. As the UQ approaches applied in this thesis work with the gPC effectively as it will be demonstrated throughout the thesis, the gPC is extensively applied in this thesis to solve the computational issue.

In conclusion, the framework for analysis and identification using Bayesian UQ has a following overview.

- The background knowledge of model users is exploited by formulating the incomplete knowledge in form of the uncertainties in the model parameter in the Bayesian sense.
- The uncertain parameters are described as random variables, whose probability distribution functions can be derived by the maximum entropy principle. In case of DPSs the uncertainties can be described by random fields.
- The randomness of parameters leads to the Initial-Boundary Value Problem (IBVP) with random variables. The IBVP is solved by means of appropriate discretization methods.
- Because solving IBVP with random variables requires normally heavily computational effort, the generalized Polynomial Chaos (gPC) is applied to solve the problem.

- The coefficients of the gPC are computed by the non-intrusive spectral projection method (NISP). The problem of multidimensional integration is solved by means of the sparse grid cubatures.
- The combination of UQ methods such as the uncertainty propagation and the sensitivity analysis can support the understanding of systems via the use of gPC.
- The gPC-approximation is exploited in the parameter calibration problem by using the gPC as a surrogate model for the forward model in inverse problems.
- The Polynomial Chaos approach is applied to a linear Bayesian filter, which results in an efficient filter to estimate the parameters or the states of the system.
- All procedures collaborate in the closed loop to improve knowledge about the system in order to overcome the two aforementioned difficulties.

In the course of this thesis, the software package for the framework is developed. The software offers the gPC computation for three UQ approaches, namely the uncertainty propagation, sensitivity analysis and the Bayesian parameter calibration. The software allows different visualization tools for the model user to support understanding the system behavior. This software is also considered as one of the contributions of the thesis.

The other contribution of the thesis is the application of the framework to a real industrial system. The proposed framework is applied to an industrial glass forming process to analyze the model and calibrate the model parameters.

Publications

In the course of the thesis, the work resulted in several publications. A first sensitivity analysis study of the glass forming process model is conducted and published in [Ja11]. The sensitivity is calculated by means of local perturbations with the finite difference approximation. The statistical inverse problem formulation for the distributed parameter systems is investigated and reported in [Ja12]. The inverse uncertainty quantification is applied to the glass forming process model to estimate the model parameters in [Ja13a]. The application of the generalized polynomial chaos (gPC) to

UQ in a distributed parameter systems is investigated and studied in[Ja14]. The development of the 3D glass forming process model is presented in [Ja15b, Ja13b]. The framework concept for analysis and identification non-linear DPS using gPC and its application to the glass forming process model is presented in [Ja15a].

1.4 Thesis organization

The thesis comprises seven chapters. After the introduction in this chapter, the framework of analyzing and identification, which is the main contribution of the thesis, is proposed in **chapter 2**. It begins with the mathematical description of DPS and the modeling of uncertainties in DPS. The discretization methods, which are required to solve a system of PDEs with random variables, are then explained. Finally, the procedure of the framework of analyzing and identification DPS with probabilistic UQ is proposed at the end of the chapter.

The generalized polynomial chaos (gPC), which is the main mathematical tool in this thesis, is discussed in **chapter 3**. The theoretical fundamentals of the gPC are explained in the beginning of the chapter. The computational methods needed for the uncertainty quantification with gPC are discussed next. The software developed in the course of the thesis for the gPC computation is then presented. The chapter ends with a numerical example of uncertainty propagation with gPC compared with a conventional approach as sampling-based method.

The other main UQ approach applied in this thesis, sensitivity analysis (SA), is discussed in **chapter 4**. The chapter gives a brief introduction of methods in sensitivity analysis, which can be classified in two groups, namely a local and global sensitivity analysis. The challenges of the sensitivity analysis computation for DPSs are discussed and their solutions with gPC are also proposed with numerical examples of sensitivity analysis.

Chapter 5 present an approach for Bayesian parameter calibration. The Bayesian statistic formulation of an inverse problem and the typical batch determination of the posterior PDF are discussed at the beginning of the chapter. Then, the Bayesian updating by means of polynomial chaos expansion (PCE) is proposed next. The exploitation of the gPC as a surrogate model is then discussed. Finally, all parameter calibration approaches discussed in the chapter are demonstrated by means of a numerical example.

The application of the framework proposed in this thesis to the glass forming process model is presented in **chapter 6**. The chapter begin with elaborating the mathematical model of the glass forming model. Then the framework is applied to the 1D and 2D model to analyze the system and calibrate the model parameters with the approaches proposed in previous chapters. **Chapter 7** concludes the work and addresses some open issues for further research.

2

Comprehensive Framework for Analyzing and Identification with Uncertainty Quantification

In this chapter the comprehensive framework for analyzing and identification of distributed parameter systems (DPS) with Bayesian UQ is proposed, which is the main contribution of this thesis. The mathematical model of DPS and the modeling of uncertainties in DPS are described first. The modeling of uncertainties in DPS with Bayesian approach establishes a system of PDEs with random variables. In general, solving such systems is achieved by discretization methods, which are explained next. The procedure of the framework of analyzing and identification of DPS with probabilistic UQ is proposed at the end of the chapter.

2.1 Modeling of uncertainties in distributed parameter systems (DPSs)

Uncertainties are commonly present in modeling. The word *uncertainty* is widely used in many contexts from science and engineering to policy and management. Each field has its own definition and typology of uncertainties based on its purposes. In this thesis only the uncertainty in distributed parameter systems (DPSs) is focused.

In this section the different types of uncertainties regarding to the interpretations of probabilities in the modeling will be pointed out. This leads to the definition of the uncertainty used in the course of this thesis. The mathematical formulation of DPS and all possible sources of uncertainties in the DPS are assorted and discussed. The uncertainties in the model are casted as random variables (RV), stochastic processes, or random fields, which are described next. The methods for the assigning the probability density functions of RVs are explained at the end of the section.

2.1.1 Aleatory vs. epistemic uncertainties

Depending on the different interpretation of probabilities between the frequentist and the Bayesian, the probability of the random outcome can be interpreted either as its long-run frequency of occurrence or as a measure of its subjective uncertainty. Because of this difference, the uncertainty is categorized into the *aleatoric* and *epistemic* uncertainty in the UQ community. Aleatoric uncertainty, also called irreducible uncertainty, is uncertainty due to inherent variation or randomness. Epistemic uncertainty arises due to an incomplete knowledge on the part of the analyst conducting the modeling and simulation.

This separation of both uncertainties comes from the Bayesian versus frequentist view of probability [Jay03]. In the frequentistic framework, only quantities with aleatoric uncertainty may be represented by random variables. The pobability distribution function of random variables can be constructed from their observed inherent variability. Contrarily, there is no fundamental theory for assigning PDEs for quantities with epistemic uncertainty from the frequentist viewpoint [Naj09].

This difficulty does not arise in the Bayesian viewpoint, in which probability is naturally the degree of belief in a proposition, and it is not necessary to be derived from sampling or observation [Naj09]. In the Bayesian framework, the probability is used to express the subject's ignorance, or a lack of information. In principle, both epistemic and aleatoric uncertainties can be treated using probability theory in the Bayesian framework.

The separation of both uncertainties is, however, still useful in the sense to identify, which uncertainties can be reduced or not. Uncertainties are characterized as epistemic, if the "modeler" sees a possibility to reduce them, and as aleatoric, if they are not reducible. This is also important for the sake of transparency in decision making [Kiu09]. As it is based on the knowledge of the modeler to foresee the possibility of reducing the uncertainties, it means that under the Bayesian framework the separation of both uncertainties is abstract and depends on the modeler's available information or knowledge.

In conclusion, using the representation of probability within the Bayesian framework, probability is used to express subject's ignorance, or a lack of information of model user.

2.1.2 Modeling of knowledge

As stated that the uncertainty depends on the modeler's knowledge, the knowledge is however a very broad term and there are also numerous definitions of knowledge in different branches.

In the scope of science considered in this thesis, knowledge is for example, the acknowledgment of the physical laws, empirical assumptions, the understanding of the intrinsic relationship between model parameters of the process, the statement of the model parameter and the prior measurement.

Such knowledge can be in an exact form of equations or deterministic values, or in a vague form of inequations, intervals or statistic values like means, variances, moments or a probability distribution function. This knowledge normally comes from the expert's experiences, observations of the process and measurements. The vague form or the statistic values can be expressed as the subject's ignorance, or lack of information, which represents the epistemic uncertainty. This means that the uncertainty is linked to the knowledge of model user under the Bayesian framework.

Representing the uncertainty with the probability under Bayesian framework as the subject's ignorance is presented in literature for some applications, e.g. [Bey99] for metrology, [Rus03] for artificial intelligence and [Wal03] for decision making. In this thesis, this representation is applied to system analysis and identification of DPS systems.

Based on the definition of uncertainty in [Bey99] and [Wal03], the uncertainty in the context of this thesis should be defined as:

> *"Any deviation between the present knowledge (of the modeler) and the unachievable ideal of completely deterministic knowledge of the relevant system".*

2.1.3 Uncertainties in DPS

A **distributed parameter system** (DPS) is a system, whose spatio-temporal variability plays an important role. Such systems are common in many engineering branches e.g. biotechnology, chemical engineering or advanced process manufacturing. This spatio-temporal variability is essential for analysis and control of the processes. Conventionally, modeling of DPS relies on the assumption of the complete knowledge of the system, i.e. the equation of the system, material characteristics, and the interaction at the boundary are entirely known. This results in deterministic models, which are employed widely in engineering and sciences and discussed in following.

Mathematical description of DPS

Many physical phenomena can be described with the relation between some continuously varying quantities and their rates of change in some independent variables. This relation can be mathematically formulated in form of differential equations. If there is only a single independent variable and its derivatives in the differential equation, the equation will be called *ordinary differential equation* (ODE). On the other hand, equations involving an unknown function of several independent variables and their partial derivatives with respect to those variables, this is termed as *partial differential equation* (PDE).

In most engineering systems the independent variables are normally time and space coordinate. Mathematically, the **distributed parameter system** (DPS) is the system, whose physical quantities rely on the space coordinate. The physical quantity x, considered as system state, is determined as a function $x(\boldsymbol{r},t) : \mathcal{G} \times \mathcal{T} \rightarrow \mathscr{X}$ with $\mathcal{G} \subseteq \mathbb{R}^3, \mathcal{T} \subseteq [0,\infty), \mathscr{X} \subseteq \mathbb{R}$ of time $t \in \mathcal{T}$ and spatial coordinate $\boldsymbol{r} = [x,y,z]^T \in \mathcal{G}$. A partial differential equation (PDE) for the function $x(\boldsymbol{r},t)$ is an equation of the form

$$\mathcal{F}\left(x(\boldsymbol{r},t), s(\boldsymbol{r},t), \frac{\partial x(\boldsymbol{r},t)}{\partial t}, \ldots, \frac{\partial^k x(\boldsymbol{r},t)}{\partial t^k}, \nabla x(\boldsymbol{r},t), \ldots, \nabla^j x(\boldsymbol{r},t)\right) = 0, \tag{2.1}$$

where $\mathcal{F}(\cdot)$ is the function relating the state $x(\boldsymbol{r},t)$ and its derivatives. $\nabla^j = \frac{\partial^j}{\partial x^j} + \frac{\partial^j}{\partial y^j} + \frac{\partial^j}{\partial z^j}$ denotes a derivative respect to space coordinate and $s(\boldsymbol{r},t)$ denotes inhomogeneous term. Additional conditions for the PDE are generally required in order to be solved for a physical system. Based on physical systems, the additional condition can be usually classified in two types relating to the independent variable, namely the initial condition relating to time and the boundary condition relating to space. The initial condition (IC) can be formulated in general form

$$\frac{\partial^i}{\partial t^i} x(\boldsymbol{r}, t=0) = g_i(\boldsymbol{r}), \qquad \text{for } i = 0, \ldots, k-1. \tag{2.2}$$

The functions $g_i(\boldsymbol{r})$ describe the relation of the state $x(\boldsymbol{r},t)$ for all spatial coordinate $\boldsymbol{r} \in [x,y,z]^T$ at the initial time $(t=0)$.

The interaction between the system and the environment is formulated in form of the boundary conditions (BC). There are many types of BC, of which only two most relevant BCs are considered in this thesis. These are

$$\text{Dirichlet type} \qquad x\left(\boldsymbol{r} \in \partial\mathcal{G}^D, t\right) = b^D(t), \tag{2.3a}$$

$$\text{Neumann type} \qquad \boldsymbol{n} \cdot \nabla x\left(\boldsymbol{r} \in \partial\mathcal{G}^N, t\right) = -b^N(t). \tag{2.3b}$$

The functions $b^D(t)$ and $b^N(t)$ describe the relation of the state $x(\boldsymbol{r},t)$ at boundary's spatial coordinates $\boldsymbol{r} \in \partial\mathcal{G}$ for all time t.

The PDE (Eq. (2.1)), ICs (Eq. (2.2)), and BCs (Eq. (2.3)) establish an **initial-boundary value problem (IBVP)**, which is used as a mathematical model to describe physical phenomena of DPS in engineering and sciences. All the parameters in these equations are collected in the parameter vector $\boldsymbol{q}$.

Defining an operator $\mathcal{M}_s(\cdot)$ containing the PDE, IC and BCs, the distributed properties and dynamic behaviors of the DPS is the solution of *system equation*

$$\mathcal{M}_s\left(x(\boldsymbol{r},t), \boldsymbol{q}\right) = 0. \tag{2.4}$$

From the system theory point of view only some values can be observed in principle, e.g. sensor measurements at some points and some sampled moments in time. These observed values are defined as the system response or system output. The mathematical interpretation of the system response is an operator on the system state and on the measurement noise w as

$$\boldsymbol{y} = \mathcal{M}_m\left(x(\boldsymbol{r},t), w\right). \tag{2.5}$$

Convection-diffusion system

The framework presented in this thesis can be applied to general space-time continuous systems, namely to initial-boundary value problems in the mathematical sense. Regarding the application problem discussed in chapter 6, a convection-diffusion system is explicitly discussed in this thesis. The convection-diffusion system describes physical phenomena where physical quantities are transferred inside a physical system due to the diffusion and convection process. The system occurs in many application in engineering and science. This system is derived normally from the conservation laws of a particular measurable physical quantities. The convection-diffusion system usually contains first-order partial derivatives with respect to time and maximal second-order with respect to space. The convection-diffusion system has a general form in spatial description as

$$\frac{\partial}{\partial t} x(\boldsymbol{r},t) = \nabla \cdot (\nu \nabla x(\boldsymbol{r},t)) - \nabla \cdot (\boldsymbol{v} x(\boldsymbol{r},t)) + \boldsymbol{s}\,(x(\boldsymbol{r},t)), \tag{2.6}$$

where

- x is the physical quantity,
- ν is the diffusion coefficient or heat conductivity,
- $\boldsymbol{v} = [v_x \; v_y \; v_z]^T$ is the velocity ,
- s describes the source of the quantity x . .

Depending on the physical quantity x, the convection-diffusion equation (2.6) coins out many well known equations. The principle of conservation of mass results in the continuity equation. In case of $x = \rho v_x$, ρv_y or ρv_z, where ρ denotes the density of fluid, these are the principles of balance of linear momentum which leads to the *Navier-Stokes* equation. By using the relation in the thermodynamics, the conservation of the energy results a heat transfer equation, which is broadly applied in engineering and sciences.

Sources of uncertainty in DPS

However, achieving the complete knowledge about the system is very difficult or even impossible in most modeling cases. Moreover, all systems possess inherent variations or randomness in the nature. Regarding the mathematical description of DPS discussed previously, the uncertainties in DPS in this thesis are categorized by sources as follows:

- *Uncertainties in the model structure*
  A model is an abstraction of the system of interest. Partial differential equations (2.1) contain some necessary assumptions. In addition, in order to keep the model computable, many simplifications like linearization of nonlinear models are necessary. Sometimes the system of interest is so complex that no exact structure for a model is available. Therefore, the model structure uncertainty always exists, even if all parameters in the model are exactly known. This model structure uncertainty is related to the function $\mathcal{F}$ in equation (2.1) and can be modeled as the systematic measurement error in eq. (2.5).

- *Boundary and Initial condition uncertainties*
  The system of interest has to be separated from its environment. The identification of the system boundaries is not a simple task in general. Interaction between the external excitations and the system state at the boundary is often neglected. Both, the system boundary in the spatial and in the time domain and the initial state of the model are difficult to measure or determine exactly. This leads to the uncertainty in the boundary condition eq. (2.2) and the initial condition eq. (2.3).

- *Algorithmic uncertainties*
  The IBVP is usually solved with numerical methods such as Finite Difference Method (FDM) or Finite Element Method (FEM) implemented in a computer. Due to numerical approximations and round-off errors, the solution of the model may have discrepancies from the true values. Moreover, the computational code should be algorithmically verified, whether the numerical model corresponds to the underlying mathematical model. The numerical analysis theory can be applied to quantify the inherent numerical errors of the computational model due to the approximation. It is important to distinguish between uncertainty in the modeling and numerical errors in the computation. The code verification and the assessment of the numerical errors are important but outside the scope of this thesis. Therefore, in this thesis, it is assumed that the numerical model of the considered system provides no numerical errors. For more detail about the numerical uncertainty and model verification, the reader can see the [Roa97], [Obe98a] and the reference therein.

- *Measurement inaccuracies*
  It is known that measurement errors are always present in measurements. There are two types of measurement errors, the systematic errors and random errors. The expression and quantification of measurement uncertainties are standardized and can be founded in numerous metrology literature. The classical statistics is a relevant mathematics tool to express the measurement uncertainties. The fundamentals about the measurement uncertainties can be found in [Pes03]. Regarding the mathematical model used in this thesis, the measurement uncertainties can be integrated in the model output equation (2.5) as noise model.

- *Parameter uncertainties*
  The parameter vector $\boldsymbol{q}$ in the equation (2.4) determines the behavior of the state and the model outputs. Often it is difficult to determine the exact values of these parameters, especially in distributed parameter system, where the parameters could be inhomogeneous in the spatial domain $\mathcal{G}$.

Although they are many kinds of uncertainties arising in the modeling and simulation of DPSs, all uncertainties discussed above can be parametrized and interpreted in form of parameter uncertainties in general [Naj09]. Therefore, only the parameter uncertainties are concerned through the thesis.

2.1.4 Probabilistic description of parameter uncertainties in DPS

Parameters in the parameter vector $\boldsymbol{q}$ could possess uncertainties resulted by the ignorance or a lack of information. Using the representation of probability with in the Bayesian formalism, the parameter with uncertainties can be represented by vector of random variables (RVs) $\boldsymbol{Q}(\omega)$. The definition of RV can be found in the appendix of the thesis.

In the systems considered in science and engineering, the variables are normally functions of space and/or time. In order to quantify the uncertainty of the system with the probabilistic approach, a representation of the randomness as function of space and/or time is necessary. Mathematical models describing this are *stochastic process* or *stochastic field*. These terms are also equivalent to the terms *random process* or *random field* respectively.

The stochastic process and random field can be seen as the extension of the notion of random variable and random vector that incorporate a dependence on time or/and space coordinates. Matthies [Mat07] proposed a representation of the stochastic process in the context of functional analysis. There, the random vector defined by the $\mathbb{R}^d$-valued RV, which is the mapping $\boldsymbol{X} : (\Omega) \to \mathbb{R}^d$, can be generalized to the $\mathcal{V}$−**valued random variable**, denoted by $X_{\mathcal{V}}$. (see definition in appendix A)

The space of all $\mathcal{V}$-valued RVs can be considered as linear combinations of elements $X(\omega)\boldsymbol{v}$, where $X(\omega)$ is real-valued RV, and $\boldsymbol{v} \in \mathcal{V}$. In other words, the $\mathcal{V}$-valued RVs are elements of tensor product $\mathcal{L}_2(\Omega) \otimes \mathcal{V}$, where $\mathcal{L}_2(\Omega)$ is the space of real-valued random variables.

Considering the vector space $\mathcal{V}$ as a space of functions $\mathcal{F}(\mathcal{T})$ on the interval $\mathcal{T}$, the $\mathcal{F}(\mathcal{T})$-valued random variable $X_{\mathcal{F}(\mathcal{T})}$ is a function $\mathcal{X} : \mathcal{T} \times \Omega \to \mathcal{F}(\mathcal{T})$ with argument -time- $t \in \mathcal{T}$ and -event- $\omega \in \Omega$. The function $\mathcal{X}(t;\omega) \in \mathcal{L}_2(\omega) \otimes \mathcal{F}(\mathcal{T})$ is then called a **stochastic process**.

The stochastic process can be treated as functions of one deterministic argument (in most cases regarded as time) whose values are random variables. The description of random fields $\mathcal{X}(\boldsymbol{r};\omega)$, where the random variable is assigned to each point $\boldsymbol{r} \in \mathcal{G}$ in the spatial domain $\mathcal{G}$, is similar to the context of stochastic processes. The concept of stochastic processes should be generalized so that the underlying argument can be multidimensional vectors or points on a manifold. Instead of one-dimensional interval $\mathcal{T}$, the multidimensional domain like spatial domain $\mathcal{G}$ can be also considered. In case that the space $\mathcal{V}$ is a space of functions $\mathcal{F}(\mathcal{G})$, the function $\mathcal{X}(\boldsymbol{r};\omega) \in \mathcal{L}_2(\omega) \otimes \mathcal{F}(\mathcal{G}) : \mathcal{G} \times \Omega \to \mathcal{F}(\mathcal{G})$ is termed as a **random field**.

There are apparent differences between a stochastic process and a random field. The stochastic process possesses the causality property resulting from the oriented nature of time, whereas the random field does not. However this difference will not be considered in this thesis, and the term *random field* will be applied through the thesis.

The formulation of the stochastic spectral approach utilizes the properties of the Hilbert space to develop an approximation. In order to use the stochastic spectral method, the random variables must have a finite variance. Therefore, in the thesis all the stochastic quantities considered are in $\mathcal{L}_2$. More information about probability theory, the notation and the definition used in this thesis can be found in the appendix.

2.1.5 Assigning probability distribution functions

Determining the uncertainties as RV means that some probability distributions have to be assigned to the $\boldsymbol{Q}(\omega)$. The probability density function (pdf) should reflect the state of knowledge about the uncertain parameters. In case of a random field, one has to discretize the random field into a random vector by separating between the stochastic space and/or spatial space or time space (see section 2.2.3).

There are several principles to derive a pdf from the available information. The methods can be found in Bayesian statistic literature under the topic of assigning the prior distribution e.g. [vdL14, Gel14]. The established methods are, for example the transformation invariance principle, the reference a prior [Ber79] or the maximum entropy principle [Jay57, Jay68]. The *Maximum Entropy Principle* is considered as an efficient method allowing to construct probability distributions from available information. In the developed framework in this thesis the background knowledge is exploited using the Maximum Entropy Principle to determine the pdfs or RVs. The reader is referred to [Jay68, vdL14, Gel14] for more information about assigning probability distributions.

Maximum Entropy Principle

The *Maximum Entropy Principle* (MaxEnt) [Jay57, Jay68] is a rule for converting a certain type of information, called testable information, to a probability distribution function. The available knowledge is called testable information, if it is possible to verify whether the derived probability distribution fulfills the constraint by the information or not. Throughout this thesis, testable information shall be always assumed, as it should not be a significant limitation for available knowledge in practice [Bey99]. The principle allows an objective representation of available testable information based on representing the remaining ignorance with least bias in the context of statistical inference.

The Maximum Entropy Principle is as follows: Given some partial information about the random variable, the probability distribution for the RV is chosen, so that it is consistent with the given information, but has otherwise the largest entropy.

The entropy of information of a continuous random variable ξ, is referred to as *differential entropy*, is defined by

$$h(\xi) = -\int_{-\infty}^{+\infty} p(\xi)\log(p(\xi))d\xi. \tag{2.7}$$

The pdf can be obtained by solving the optimization problem

$$p(\xi) = \underset{p \in \mathcal{C}_p}{\arg\max}\, h(\xi), \tag{2.8}$$

where $\mathcal{C}_p$ is the admissible set of all pdfs satisfying the constraints defined by the available information. The construction of pdfs for random fields or stochastic processes using the Maximum Entropy Principle is discussed in [Soi08]. In the paper, the MaxEnt optimization problem (2.8) is solved by the Lagrange multiplier associated with the constraints. In general, engineers or scientists possess typical information such as the interval, the mean or the standard deviation. For most typical information found in engineering, the maximum entropy pdfs of RVs are the following:

- If nothing is known besides the mean μ and the standard deviation σ, the distribution of RV is **normal** $\mathcal{N}(\mu,\sigma^2)$.

$$p_\xi(\xi|\mu,\sigma) = \frac{1}{\sigma\sqrt{2\pi}} \exp\left(-\frac{(\xi-\mu)^2}{2\sigma^2}\right)$$

- If the interval of the RV $[a,b]$ is known, the **uniform distribution** on the interval $[a,b]$ is the maximum entropy distribution among all continuous distributions supported in the interval $[a,b]$

$$p_\xi(\xi|a,b) = \begin{cases} \frac{1}{b-a}, & \text{for } a < \xi < b, \\ 0, & \text{otherwise} \end{cases}$$

- If the mean value of parameter is known and the parameter is positive. **The exponential distribution**

$$p_\xi(\xi|\mu > 0) = \begin{cases} \frac{1}{\mu}\exp(\frac{-\xi}{\mu}), & \text{for } \xi \geq 0 \\ 0, & \text{for } \xi < 0 \end{cases}$$

  is the distribution with the maximum entropy among all continuous distributions with support $[0,\infty)$ that have a specified mean of μ.

Applications of MaxEnt can be found in many fields e.g. thermodynamics, statistic mechanics, signal processing, and pattern recognition. Further elaborate information and pdfs regarding to other applications can be found in [Kap93].

2.2 Partial differential equations with random variables and its discretizations

Considering a distributed parameter system with the model $\mathcal{M}_s\left(x(\boldsymbol{r},t),\boldsymbol{q}\right)$ as mentioned in the section 2.1.3, where $x(\boldsymbol{r},t)$ is the state of the system at the spatial coordinate $\boldsymbol{r}$ at time t and $\boldsymbol{q}$ is vector of parameters of the model. As discussed in the last section the knowledge about the model of the user can be transformed into uncertainties of the model. The uncertainties of the model are described by representing the vector parameter $\boldsymbol{q}$ as random variable or random field, whose joint probability density $p(\boldsymbol{q})$ can be determined by the *MaxEnt.* The randomness of the parameters leads to a PDE with random variables. It is noted that a PDE with RVs is not the same as a *stochastic differential equation.* In many papers, these terms are sometimes treated as synonymous, although in fact they require different mathematical techniques to solve.

A *stochastic differential equation* (SDE) is an equation, which possesses stochastic differential terms. The solution of a SDE is a stochastic process and exhibits non-differentiable sample paths. Therefore, these equations require the Ito or Stratonovic calculus to solve [Smi13]. An example for SDE is the *Langevin Equation.* It is noted that SDEs are out of scope of this thesis and therefore are not further discussed. For further details about SDEs, the interested reader is referred to [Gar88] and [Klo92].

In contrast concerning PDEs with RVs, the randomness is manifested in the parameters which are continuous with respect to time and space. For each realization of $\omega \in \Omega$, a random differential equation is analyzed and solved by using the theory of standard differential equations. The solutions of the partial differential equation with random variables are therefore collections of the smooth sample paths of all realizations. Only PDEs with RVs are considered in the course of this thesis.

The probabilistic UQ of DPS needs the computation of the PDE for all realizations. Because there are no readily available solution for such PDEs in general, the PDEs are usually solved by numerical methods, which solve the PDE as finite dimensional problems. The approximation methods from infinite-dimension to finite-dimension for spatial, time and stochastic space are discussed in this section.

2.2.1 Deterministic spatial discretization

The initial-boundary value problem (IBVP) is usually only in some cases analytically solvable. The IBVP is practically approximated from the infinite-dimensional problem to a finite dimensional problem. This approximation is usually applied in spatial domain and can be accomplished by transforming the PDE and the boundary conditions into a system of ordinary differential equations (ODEs). An overview of important spatial discretization methods can be founded in [Li11].

As shown in figure 2.1 the spatial discretization methods can be classified in two groups.

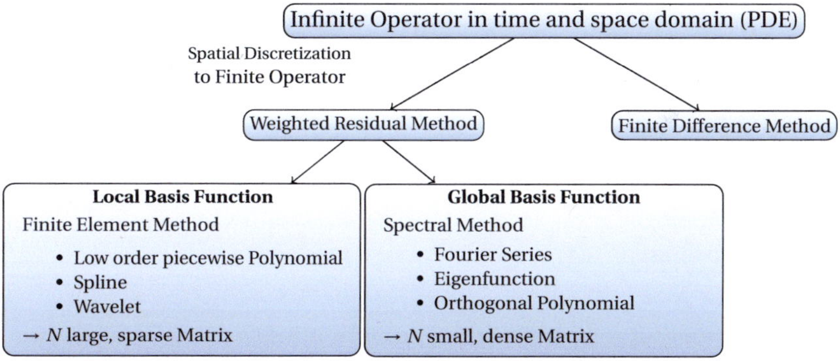

Figure 2.1: Classification of discretization methods to approximate a PDE into system of ODEs

- **Finite Difference Method** (FDM): this method is an approximation of the differential operator in the PDE by difference operator as

$$\frac{\partial x}{\partial r} \approx \frac{\Delta x}{\Delta r}. \tag{2.9}$$

  This approximation term can be a forward, a backward or a central difference often derived from a Taylor expansion.

- **Weighted Residual Method** (WRM): It is known that a continuous function can be represented as infinite series, e.g. Fourier expansion. Based on this principle, the spatial continuous solution of IBVP at any time t can be expanded by a set of spatial basis functions $\{\phi_i(\boldsymbol{r})\}_{i=1}^{\infty}$ [Li11] as

$$x(\boldsymbol{r},t) = \sum_{i=1}^{\infty} x_i(t)\phi_i(\boldsymbol{r}). \tag{2.10}$$

From this infinite series the solution of IBVP can be approximated in finite dimension by truncating the infinite series at N_s

$$x(\boldsymbol{r},t) \approx \hat{x}(\boldsymbol{r},t) = \sum_{i=1}^{N_s} x_i(t)\phi_i(\boldsymbol{r}) := \boldsymbol{\phi}(\boldsymbol{r}) \cdot \mathbf{x}(t), \tag{2.11}$$

where $\boldsymbol{\phi}(\boldsymbol{r}) = \left[\phi_1(\boldsymbol{r}),\ldots,\phi_{N_s}(\boldsymbol{r})\right]^T$ denotes the vector of shape functions and $\mathbf{x} = \left[x_1(t),\ldots,x_{N_s}(t)\right]^T$ represents corresponding coefficients. Let the solutions $x(\boldsymbol{r},t)$ belong to an appropriate deterministic space V and $V^{N_s} = \operatorname{span}\{\phi_i\}_{i=1}^{N_s} \subset V$. The finite solution $\hat{x}(\boldsymbol{r},t)$ is the approximation of the solution $x(\boldsymbol{r},t)$ on the subspace V^{N_s}.

This approximation has a residual between the exact solution and the approximation as:

$$e = \mathcal{F}\left(\hat{x}(\boldsymbol{r},t)|\boldsymbol{q}\right) \tag{2.12}$$

The philosophy of the WRM is based on obtaining the approximate solution which assumes the residual to be zero over the domain. This can be achieved by introducing *weighting functions*, also called *test functions*, $\{\varphi(\boldsymbol{r})\}_{i=1}^{N_s}$ so that the inner products vanish.

$$\left\langle e|\varphi_i\right\rangle = 0, \qquad \text{for } i = 1,\ldots,N_s \tag{2.13}$$

This formulation is called *weak formulation.* The solution of the weak formulation is also only an approximate solution with respect to certain selected test functions.

This method is commonly employed and very efficient. Based on the types of spatial basis functions, either local or global, WRM can be classified in two groups namely the *Finite Element Method* and *Spectral Method.* The Spectral Methods have advantageous error properties with exponential convergence, if the solution is smooth. Compared to the local basis functions, by which convergence rate stays constant regardless of the function smoothness.

In both methods, proper spatial basis functions can be used to expand the solution into finite series. The selection of the basis functions is a critical point and has always an impact to the performance of solving system.

It is noted that the FDM is based on an approximation of the derivative operators of the PDE, while the WRM approximates the solution of the PDE. However the FDM can be classified as a Finite Element Method approach, in case that the spatial basis functions and the weighting functions are both chosen as *Dirac Delta functions* [Li11].

Finite Element Method

The *Finite Element Method* (FEM) is a numerical technique to approximate solutions of boundary value problems (BVPs). The FEM is also one of the most applied methods to discretize the spatial domain of PDE. As mentioned before, the FEM can be classified as a weighted residual method (WRM), in which the spatial basis functions $\{\phi(\boldsymbol{r})\}$ are chosen to be local basis functions. The spatial domain is discretized into many small sub-domains, called finite elements, in general with the help of a *low-order piecewise polynomial* function. But other local functions, e.g. *wavelets* [Ko95, Mah00] or *splines* [Höl03], are also possible. It is also noted that wavelets can be considered as global basis function as well.

In general in FEM, the test functions $\{\varphi(\boldsymbol{r})\}_{j=1}^{N_s}$ are often selected to be the spatial basis functions $\{\phi(\boldsymbol{r})\}_{i=1}^{N_s}$. The method using spatial basis functions as test functions is known as *Galerkin Method.* The residual is orthogonal to all test function in V^{N_s} yielding

$$\int_{\mathcal{G}} \phi_j(\boldsymbol{r}) \left(\mathcal{F}\left(\hat{x}(\boldsymbol{r},t)|\boldsymbol{q}\right)\right) d\boldsymbol{r} = 0 \qquad \text{for } j = 1,\ldots,N_s \qquad (2.14)$$

As the residual is orthogonal to each basis function, therefore, the best solution possible in the space $V^{N_s} = \operatorname{span}\{\phi_j\}_{j=1}^{N_s} \subset V$.

Spatial discretization of the diffusion-convection system

As an example for the FEM formulation the diffusion-convection system of section (2.1.3) is presented and discussed in the following. The FEM formulation starts with the weak formulation of the diffusion-convection system equation (2.6), it is

$$\int_{\mathcal{G}} \varphi_j(\boldsymbol{r}) \left(\frac{\partial}{\partial t} \hat{x}(\boldsymbol{r},t) - \nabla \cdot (\nu \nabla \hat{x}(\boldsymbol{r},t)) + \nabla \cdot (\boldsymbol{v} \hat{x}(\boldsymbol{r},t)) - s(\boldsymbol{r},t) \right) d\boldsymbol{r} = 0, \quad (2.15)$$

for $j = 1,\ldots,N_s$. By means of the additive property the weighted integral over the spatial domain $\mathcal{G}$ can be rewritten as follows:

$$\int_{\mathcal{G}} \varphi_j \frac{\partial \hat{x}}{\partial t} d\boldsymbol{r} - \int_{\mathcal{G}} \varphi_j (\nabla \cdot (\nu \nabla \hat{x}))\, d\boldsymbol{r} + \int_{\mathcal{G}} \varphi_j (\nabla \cdot (\boldsymbol{v} \hat{x}))\, d\boldsymbol{r} - \int_{\mathcal{G}} \varphi_j s d\boldsymbol{r} = 0. \quad (2.16)$$

The integral of the diffusion term can be reformulated by using partial integration.

$$\int_{\mathcal{G}} \varphi_j (\nabla \cdot (\nu \nabla \hat{x}))\, d\boldsymbol{r} = \int_{\mathcal{G}} \left(\nabla \cdot (\varphi_j \nu \nabla \hat{x})\right) d\boldsymbol{r} - \int_{\mathcal{G}} \left(\nabla \varphi_j\right) \cdot (\nu \nabla \hat{x})\, d\boldsymbol{r} \quad (2.17)$$

The integral term can further manipulated by using the *Gauss' divergence theorem* given as follows:

$$\int_{\mathcal{G}} \left(\nabla \cdot (\varphi_j \nu \nabla \hat{x})\right) d\boldsymbol{r} = \oint_{\partial \mathcal{G}} \varphi_j \underbrace{\boldsymbol{n}^T \nabla \hat{x}}_{=:-b^N(\boldsymbol{r},t)} dA \quad (2.18)$$

Using equations (2.17) and (2.18), the equation(2.16) can be rewritten as:

$$\int_{\mathcal{G}} \varphi_j \frac{\partial \hat{x}}{\partial t} d\boldsymbol{r} + \int_{\mathcal{G}} \left(\nabla \varphi_j\right) \cdot (\nu \nabla \hat{x})\, d\boldsymbol{r} + \int_{\mathcal{G}} \varphi_j (\nabla \cdot (\boldsymbol{v} \hat{x}))\, d\boldsymbol{r}$$
$$= \int_{\mathcal{G}} \varphi_j s d\boldsymbol{r} - \oint_{\partial \mathcal{G}} \varphi_j b^N dA \quad (2.19)$$

Applying the *Galerkin Method* to equation (2.19) by substituting the test functions $\{\varphi_j(\boldsymbol{r})\}_{j=1}^{N_s}$ with the spatial basis functions $\{\phi_j(\boldsymbol{r})\}_{j=1}^{N_s}$ leads to:

$$\int_{\mathcal{G}} \phi_j \frac{\partial}{\partial t} \sum_{i=1}^{N_s} x_i(t)\phi_i d\boldsymbol{r} + \int_{\mathcal{G}} (\nabla\phi_j) \cdot \left(\nu\nabla \sum_{i=1}^{N_s} x_i(t)\phi_i \right) d\boldsymbol{r}$$
$$+ \int_{\mathcal{G}} \phi_j \left(\nabla \cdot \left(\boldsymbol{v} \sum_{i=1}^{N_s} x_i(t)\phi_i \right)\right) d\boldsymbol{r} = \int_{\mathcal{G}} \phi_j s d\boldsymbol{r} - \oint_{\partial\mathcal{G}} \phi_j b^N dA \tag{2.20}$$

This is a system of N_s-ordinary differential equation, which can be rewritten by using vector notation:

$$\int_{\mathcal{G}} \boldsymbol{\phi}(\boldsymbol{r})\boldsymbol{\phi}^T(\boldsymbol{r})\, d\boldsymbol{r} \frac{d}{dt}\mathbf{x}(t) + \int_{\mathcal{G}} (\nabla\boldsymbol{\phi}(\boldsymbol{r})) \cdot \left(\nu\nabla\boldsymbol{\phi}^T(\boldsymbol{r})\right) d\boldsymbol{r}\, \mathbf{x}(t)$$
$$+ \int_{\mathcal{G}} \boldsymbol{\phi}(\boldsymbol{r}) \left(\nabla \cdot \left(\boldsymbol{v}\, \boldsymbol{\phi}^T(\boldsymbol{r})\right)\right) d\boldsymbol{r}\, \mathbf{x}(t)$$
$$= \int_{\mathcal{G}} \boldsymbol{\phi}(\boldsymbol{r})\, s d\boldsymbol{r} - \oint_{\partial\mathcal{G}} \boldsymbol{\phi}(\boldsymbol{r})\, b^N dA \tag{2.21}$$

The equation (2.21) can be rewritten in matrix form as:

$$\mathbf{M}\frac{d}{dt}\mathbf{x}(t) + (\mathbf{N}+\mathbf{G})\mathbf{x}(t) = \mathbf{b}, \tag{2.22}$$

where elements of the $(N_s \times N_s)$ matrices $\mathbf{M}, \mathbf{N}, \mathbf{G}$ and the $(N_s \times 1)$ vector $\mathbf{b}$ are defined as follows:

$$[\mathbf{M}]_{ij} = \int_{\mathcal{G}} \phi_j \phi_i d\boldsymbol{r}, \qquad [\mathbf{N}]_{ij} = \int_{\mathcal{G}} (\nabla\phi_j) \cdot (\nu\nabla\phi_i)\, d\boldsymbol{r},$$
$$[\mathbf{G}]_{ij} = \int_{\mathcal{G}} \phi_j (\nabla \cdot \boldsymbol{v}\phi_i)\, d\boldsymbol{r}, \qquad \mathbf{b} = \int_{\mathcal{G}} \boldsymbol{\phi}(\boldsymbol{r})\, s d\boldsymbol{r} - \oint_{\partial\mathcal{G}} \boldsymbol{\phi}(\boldsymbol{r})\, b^N dA$$

For computation the stationary solutions $\mathbf{x}_\infty$, the derivative term $\frac{d}{dt}\mathbf{x}(t)$ is set to zero, it yields:

$$\mathbf{x}_\infty = (\mathbf{N}+\mathbf{G})^{-1}\mathbf{b}. \tag{2.23}$$

This equation (2.23) is explicit in case that the PDE is linear. In case of nonlinear PDE, the matrices **N** and **G** depend on **x** and the equation (2.23) require specific nonlinear solvers, which is mostly iterative.

For the time dependent solution the derivation term $\frac{d}{dt}\mathbf{x}(t)$ must be approximated, which is discussed in the following subsection.

2.2.2 Time discretization

For the time dependent problem, the solution $\mathbf{x}(t)$ is also the function of time. For the numerical solution of initial value problems, the time domain is discretized for the computational purposes. In the most simple method, the time interval $\mathcal{T} = [0, t_{\text{end}}]$ is divided into N_t equal time increments such that $t_j = j\Delta t,\ j = 0,\ldots,N_t$ where $\Delta t = t_{\text{end}}/N_t$ is the time step. The solution of the time dependent problem $\mathbf{x}(t)$ is represented by the series $\mathbf{x}(t_j)$.

To discuss about the time discretization, equation (2.22) is rewritten into:

$$\frac{d}{dt}\mathbf{x}(t) = \mathbf{M}^{-1}\left(\mathbf{b} - (\mathbf{N}+\mathbf{G})\mathbf{x}(t)\right), \tag{2.24}$$

$$\dot{\mathbf{x}}(t) = F(\mathbf{x}(t)). \tag{2.25}$$

The time derivative $\dot{\mathbf{x}}(t)$ can be approximated by the *Finite Difference*

$$\dot{\mathbf{x}}(t) \approx \frac{\mathbf{x}(t_{j+1}) - \mathbf{x}(t_j)}{\Delta t}. \tag{2.26}$$

The state $\mathbf{x}(t_{j+1})$ can be then be solved for example by inserting the state $\mathbf{x}(t_j)$ in $F(\cdot)$:

$$\frac{\mathbf{x}(t_{j+1}) - \mathbf{x}(t_j)}{\Delta t} = F(\mathbf{x}(t_j)), \tag{2.27}$$

$$\mathbf{x}(t_{j+1}) = \mathbf{x}(t_j) + \Delta t F(\mathbf{x}(t_j)). \tag{2.28}$$

This is well known as *Explicit Euler scheme.* Solving with Explicit Euler is simple but quite slow, because the time step is limited by the stability of the numerical solver regarding to the *Courant-Friedrichs-Lewy (CFL) condition* [Cou28]. Another numerical more stable alternative is inserting the state $\mathbf{x}(t_{j+1})$ in $F(\cdot)$, which is known as *Implicit Euler scheme.*

$$\dot{\mathbf{x}}(t) \approx \frac{\mathbf{x}(t_{j+1}) - \mathbf{x}(t_j)}{\Delta t} = F(\mathbf{x}(t_{j+1})) \tag{2.29}$$

The implicit approach needs an iterative solver to solve the equation, but it is unconditionally stable and can be computed with the large time steps Δt. In most of the commercial software, more efficient numerical methods for solving first-order initial value problems are employed. The methods can be classified into two categories, namely *linear multistep* methods and *Runge-Kutta* methods.

The **linear multistep** methods use several previous points and derivative values to determine the current value, whereas the **Runge-Kutta** methods take some intermediate steps to obtain the current value. In practice, the methods are chosen according to the order of accuracy and to the type of the system (stiff or nonstiff). Examples for numerical methods for time discretization are the *Dormand-Prince* method[1], the *Bogacki-Shampine method*[1], *Adams-Bashforth-Moulton* method, *Backward Differentiation Formulas* (BDFs), etc. The reader is referred to [Asc98] and [Sha97] for more information about this topic.

2.2.3 Stochastic discretization

In case of PDE systems with deterministic parameters, the time and space discretizations are employed for numerical computations as presented in previous subsections. With the probabilistic UQ formulation for PDEs, the discretization of random fields is necessary for the computation. The stochastic discretization is an approximation of a random field $\mathcal{X}(\boldsymbol{r},\omega)$ by means of a *finite set* of random variables $\{\xi_1,\ldots,\xi_{N_s}\}$ as random vectors.

As an example, a stochastic process, which represents randomness as a function of time, can be discretized by representing the distribution of a stochastic process with the collection of finite dimensional distributions of random variables $\left(X(t_1),\ldots,X\left(t_j\right)\right)$ where $t_j \in \mathcal{T}$ is the sampling time.

In case of a multidimensional interval as e.g. spatial space, the discretization of the random field is more complex than the stochastic process. Depending on the representation, the random field can be discretized by several approaches. Analogous to the stochastic process with respect to time, one of the most simple methods is the **point discretization** method.

1 The *Dormand-Prince* method is known as explicit Runge-Kutta (4,5) as it use the fourth- and fifth-order Runge-Kutta to approximate the solution. The *Bogacki-Shampine method* is known as explicit Runge-Kutta (2,3) as it use the second- and third-order Runge-Kutta to approximate the solution. Both methods use the difference between these approximated solutions to adapt the step size.

The approximation of the random process is the collection of the random vectors at some given points $\boldsymbol{r}_i$, e.g. nodes, midpoints or integration points.

$$\xi_i(\omega) = \mathcal{X}(\boldsymbol{r}_i,\omega) \qquad \text{for } i = 1,\ldots,N_s \tag{2.30}$$

The values of the random field can be calculated approximately by the interpolation with the shape basis function $\phi_i(\boldsymbol{r})$.

$$\mathcal{X}(\boldsymbol{r},\omega) = \sum_{i=1}^{N_s} \phi_i(\boldsymbol{r})\xi_i(\omega) \tag{2.31}$$

The interpolation shape function can be determined independently from the spatial discretization shape function. Due to its simpleness this method is widely applied in practice and satisfies practical requirements well. Another method is the **spatial average method** introduced by Vanmarcke and Grigoriu [Van83], where the random variables $\xi_i(\omega)$ are weighted integrals of $\mathcal{X}(\boldsymbol{r},\omega)$ over a discretized domain Ω_i

$$\xi_i(\omega) = \int_{\Omega_i} \mathcal{X}(\boldsymbol{r},\omega)\, w(\boldsymbol{r}) d\omega \tag{2.32}$$

However the most often used is **series expansion**. The expansion approximates random fields as a sum of products of functions defined on a spatial domain and functions of random variables $\Lambda(\cdot)$.

$$\mathcal{X}(\boldsymbol{r},\omega) = \sum_{i=1}^{\infty} \Lambda(\xi_i(\omega))\phi_i(\boldsymbol{r}) \tag{2.33}$$

In case that the basis functions $\phi_i(\boldsymbol{r})$ are global basis functions in the series expansion method, it is sometimes called **spectral expansion method** [Gha03],[Kee04]. The reader is referred to [Sud00] and [Mat05] for more details about discretization methods.

With the stochastic discretization, the infinite-dimensional field can be represented by the finite-dimensional approximation. However, one must be careful when extending finite-dimensional results back to the infinite-dimensional representation.

One of the most widely applied random field discretization method from this group is the *Karhunen Loève Expansion* (KLE). Therefore, KLE deserves special attention here and is discussed in the following. Another method applied in this thesis is polynomial chaos expansion, which is discussed in the section 3.1.3.

Karhunen Loève Expansion

Karhunen Loève expansion (KLE) is one of the most applied methods to decompose a stochastic process or a random field into a stochastic part and a deterministic spatial or the temporal part. It is named after Kari Karhunen and Michel Loève, who invented it independently in the forties. KLE is a representation of a stochastic process as an infinite linear combination of orthogonal functions.

In the finite-dimensional case, where the coefficients are computed from a sample, the method is named as, namely *Principle Component Analysis (PCA), Proper Orthogonal Decomposition (POD), Karhunen Loève Transform (KLT), Empirical orthogonal functions* or *Hotelling Transform.*

Consider a random field $\mathcal{X}(\boldsymbol{r},\omega)$ with its covariance function $\mathrm{Cov}_{\mathcal{X}}(\boldsymbol{r},\boldsymbol{r}')$ (see Def. 14 and 15 in appendix). If the covariance satisfies the definition of a Mercer kernel [Mer09], which requires that the covariance is symmetric and positive-definite, then the random field $\mathcal{X}(\boldsymbol{r},\omega)$ admits the decomposition:

$$\mathcal{X}(\boldsymbol{r},\omega) = \sum_{i=0}^{\infty} \sqrt{\lambda_i}\xi_i(\omega)\, x_i(\boldsymbol{r}), \tag{2.34}$$

which separates the random field into a spatial part $x_i(\boldsymbol{r})$ and the stochastic part $\xi_i(\omega)$. Let $\lambda_0 = 1, \xi_i = 1$ and $\mathcal{X}_0 = \overline{\mathcal{X}}(\boldsymbol{r})$, the equation (2.34) can be rewritten in:

$$\mathcal{X}(\boldsymbol{r},\omega) = \overline{\mathcal{X}}(\boldsymbol{r}) + \sum_{i=1}^{\infty} \sqrt{\lambda_i}\xi_i(\omega)\, x_i(\boldsymbol{r}) \tag{2.35}$$

$(\lambda_i, x_i(\boldsymbol{r}))$ is the pair of eigenvalues and orthonormal eigenfunctions of the kernel $\mathrm{Cov}_{\mathcal{X}}(\boldsymbol{r},\boldsymbol{r}')$, which is the solution of the Fredholm equation of the second-kind:

$$\int_{\mathcal{G}} \mathrm{Cov}_{\mathcal{X}}(\boldsymbol{r},\boldsymbol{r}')x_i(\boldsymbol{r}')d\boldsymbol{r}' = \lambda_i x_i(\boldsymbol{r}), \qquad \text{for } \boldsymbol{r}_1 \in \mathcal{G}. \tag{2.36}$$

With the orthogonal property of $x_i(\boldsymbol{r})$ the RVs ξ_i are given by:

$$\xi_i(\omega) = \frac{1}{\sqrt{\lambda_i}} \int_{\mathcal{G}} \left(\mathcal{X}(\boldsymbol{r},\omega) - \overline{\mathcal{X}}(\boldsymbol{r}) \right) x_i(\boldsymbol{r}) d\boldsymbol{r}. \tag{2.37}$$

Furthermore, the RVs ξ_i are centered and uncorrelated.

$$\mathbb{E}(\xi_i) = 0 \ , \ \mathbb{E}(\xi_i \xi_k) = \delta_{ik} \tag{2.38}$$

The discretization of a random field by KLE is done by truncating eq. (2.35) to a finite number of terms N_{KLE}:

$$\mathcal{X}(\boldsymbol{r},\omega) \approx \overline{\mathcal{X}}(\boldsymbol{r}) + \sum_{i=1}^{N_{\mathrm{KLE}}} \sqrt{\lambda_i} \xi_i(\omega) \, x_i(\boldsymbol{r}), \tag{2.39}$$

such a approximation is the best one achieved in $\mathcal{L}_2(\mathcal{G} \times \Omega) \cong \mathcal{L}_2(\mathcal{G}) \otimes \mathcal{L}_2(\Omega)$ norm [Ros12a].

For the computational purpose the spatial field is in general discretized as well. For example using the FEM to discretize the spatial domain, the Karhunen Loève approximation can be rewritten in a spatial discretized form as:

$$\mathcal{X}(\boldsymbol{r},\omega) \approx \left(\overline{\boldsymbol{\mathcal{X}}} + \sum_{i=1}^{N_{\mathrm{KLE}}} \sqrt{\lambda_i} \xi_i(\omega) \mathbf{x_i} \right) \cdot \boldsymbol{\phi}(\boldsymbol{r}),$$

where $\overline{\boldsymbol{\mathcal{X}}} = \left[\overline{\mathcal{X}}^1, \ldots, \overline{\mathcal{X}}^N \right]^T$ denotes a vector of mean values related to the shape function and $\mathbf{x_i}$ denotes $\left[x_i^1, \ldots, x_i^{N_s} \right]^T$, which describes the coefficients corresponding to the spatial basis functions $\{\phi_i(\boldsymbol{r})_{i=1}^{N_s}\}$. The reader is referred to [Gha03], [LM10] for more information about the KLE.

2.3 A comprehensive framework for analyzing and identification DPS with Bayesian uncertainty quantification

Usually, engineers and scientists possess some theoretical and empirical knowledge about the process, which is used to establish a computational model, to design the control strategies and to optimize the process. The knowledge is normally based on expert experience, observations of the process and measurements. Analyzing the system with simulation generates new knowledge about the system behavior, which facilitates the understanding of the system. On the other hand, identification of the system from observations and measurements can deliver some knowledge about the system as well.

The knowledge about the system plays a main role in the proposed framework. Based on the modeler's knowledge, the DPS is modeled with an initial boundary value problem $\mathcal{M}_s\left(x(\boldsymbol{r},t),\boldsymbol{q}\right) = 0$ as described in section 2.1.3.

Under the Bayesian framework, the incomplete knowledge about the system is considered as uncertainty of the system. All uncertainties of the system are parametrized and interpreted in form of parameter uncertainties. Using a probabilistic representation, the uncertain parameters can be described as a vector of random variables $\boldsymbol{Q}(\omega)$, whose pdf $p_{\boldsymbol{Q}}(\boldsymbol{q})$ can be achieved by the conversion of the knowledge about the parameters with the *Principle of Maximum Entropy.*

The modeling of uncertainties of a DPS with the Bayesian approach establishes a system of PDEs with random variables, which is to be analyzed and identified.

The framework relies on three probabilistic UQ methods namely uncertainty propagation, sensitivity analysis and Bayesian parameter calibration. The uncertainty propagation assesses the pdf of the quantities of Interest (QoI) w.r.t. the pdf of the uncertain parameters. The uncertain parameters affect the QoI unequally. The sensitivity analysis is used to identify the influence of each uncertain parameter. The pdf of the QoI and the sensitivity of the parameters are crucial knowledge for the model user to analyze in order to understand the system behavior.

With the measurements of the real process the uncertain parameters can be calibrated using Bayesian inference. The sensitivity of the parameters can be also used to rank the parameters. The model with the calibrated parameters permits better agreement between the model prediction and the measurements from the real process.

All of these approaches are operated under the Bayesian UQ framework, by considering the pdf as the knowledge as the central point of the framework. With the uncertainty propagation (UP), the sensitivity analysis (SA) and the Bayesian inference, new knowledge about the system is gained and can be used to develop the model further.

A system of PDEs with RVs requires certain mathematical techniques to be solved. Sampling-based methods such as Monte-Carlo are generally adopted in the engineering field. The methods are straightforward by solving the model $\mathcal{M}_s\left(x(\boldsymbol{r},t),\boldsymbol{q}\right)$ for all input realizations from the distribution $p(\boldsymbol{q})$. This requires however extensive computation. To reduce the computational effort, model order reduction or surrogate model approaches can be applied to resolve the issue.

In the recent years, the *Polynomial Chaos Expansion* (PCE) approach [Wie38] has been gaining more attention in many fields. The PCE and also its generalized version, proposed by Xiu and Karniadakis [Xiu02], offer an efficient way to approximate the probability distribution functions and the statistical moments of the outputs. This method can be applied to sensitivity analysis and the Bayesian inverse problem as well.

In the framework developed in this thesis the PCE and the generalized polynomial chaos (gPC) are the main instruments to support analyzing and identification. The gPC is applied to support analyzing the system with respect to four tasks.

1. Approximation of the system response distribution concerning UP;
2. Parameter estimation via the Bayesian inverse approach;
3. Global sensitivity analysis;
4. Local sensitivity analysis.

In conclusion, the comprehensive framework for analyzing and identification is organized into six procedural steps:

1. Modeling uncertainties in the DPS model from the present knowledge under the Bayesian framework;
2. Assigning probability distribution functions to the parameters;

3. Assessing the pdf of the system response quantities of interest (QoI) by means of generalized Polynomial Chaos (gPC);

4. Conduct sensitivity analysis with generalized polynomial chaos;

5. Calibrating the model parameters based on the measurement data according to the sensitivity indices;

6. Conduct model updating with the new knowledge and, if necessary, go again to the 1. step.

The procedures of each step are explained throughout the thesis.

In this chapter, the required fundamentals for the uncertainty modeling and for assigning the pdfs to the parameters are discussed. The system with uncertainty is modeled by a system of PDEs with random variables, whereas the pdfs of the RVs represent the knowledge state of the modeler. The uncertainty propagation analysis and gPC are discussed in chapter 3. The other two crucial steps, namely sensitivity analysis and the Bayesian parameter calibration, are discussed in the chapter 4 and 5 respectively.

3

Uncertainty Propagation Analysis and Generalized Polynomial Chaos

Uncertainty propagation is the study of the model outputs' response with respect to the probability distribution of the inputs. The targets of the uncertainty propagation analysis are usually:

1. Evaluating the reliability of the system responses or outputs;
2. Evaluating the statistic values like means, variances or moments of the outputs;
3. Assessing the complete probability distribution of the outputs.

This information is useful for the model user to characterize the system. Many probabilistic UQ approaches are proposed for accomplishing these three targets. Lee [Lee09] compared various UQ methods and classified them into five categories:

1. Simulation-based methods: Monte Carlo simulations, importance sampling, adaptive sampling [Mad06, Buc88] etc. ;
2. Local expansion-based methods: Taylor series, perturbation method [Mad06, Gha03, Cac03] etc. ;

3. Most probable point (MPP)-based methods: first-order reliability method (FORM) and second-order reliability method (SORM) [Has74, Fie79];

4. Functional expansion-based methods: Karhunen-Loève expansion (KLE), Neumann expansion, polynomial chaos expansion (PCE), generalized polynomial chaos expansion (gPCE) [Gha03, Xiu02, Xiu10] etc.;

5. Numerical integration-based methods: Full factorial numerical integration (FFNI) and dimension reduction (DR) [Seo02, Xu04].

The methods in each category are developed for divergent objectives and therefore all have some limitations. The most probable point-based methods are developed intentionally to evaluate the reliability of the system response. The numerical integration-based methods can evaluate merely the statistical moments but not the pdf. The local expansion-based methods has limited usage to highly nonlinear systems in general.

The pdf of the system responses is the cornerstone in the developed framework for analyzing and identification in this thesis. The pdf offers comprehensive information about the system, however, it is quite difficult to obtain. The pdf of the system response can be assessed by the simulation-based methods and the functional expansion-based methods. The simulation-based methods, such as Monte-Carlo, are generally adopted in the engineering field. The methods are straightforward by solving the model $\mathcal{M}_s\left(x(\boldsymbol{r},t),\boldsymbol{q}\right)$ for all input realizations from the distribution $p_{\boldsymbol{Q}}(\boldsymbol{q})$. This requires however extensive computation. For reducing the computational effort, approaches such as model order reduction or using a surrogate model, can be applied to resolve the issue.

In the recent years, the PCE approach has been gaining more attention in many fields. The PCE and also its generalized version, proposed by Xiu and Karniadakis [Xiu02], offer an efficient way to approximate the probability density functions and the statistical moments of the outputs. This chapter introduces the basic concept of gPC and its application to the uncertainty propagation task.

3.1 Polynomial expansions of stochastic quantities

In 1938, Norbert Wiener introduced the *Polynomial Chaos* (PC) to represent Gaussian processes by using a series of Hermite polynomials [Wie38]. It should be noted that the word *Chaos*, introduced by N. Wiener, has nothing to do with the modern term *Chaos* in mathematics where it characterizes the unpredictable behavior of dynamical systems. Theoretically, the polynomial chaos can be considered as a functional approximation of a given random variable $X(\boldsymbol{\xi}(\omega))$ by basis functions of other RVs,

$$X(\boldsymbol{\xi}(\omega)) = \sum_{k=0}^{\infty} x_k \Xi_k(\boldsymbol{\xi}(\omega)) \approx \sum_{k=0}^{N_P} x_k \Xi_k(\boldsymbol{\xi}(\omega)), \tag{3.1}$$

where $\boldsymbol{\xi}(\omega)$ is the random variable vector with known pdf $p_{\boldsymbol{\xi}}(\boldsymbol{\xi})$, $\Xi_k(\cdot)$ are suitably selected functionals of the RVs and the coefficients x_k will be called the k−th "stochastic mode"[1] of $X(\boldsymbol{\xi}(\omega))$ [LM10]. If the function $X(\boldsymbol{\xi}(\omega))$ is smooth, such an approximation possesses a *spectral convergence* property for some series of basis functions, such as series of orthogonal polynomials in PC. It means that the approximation error becomes smaller for the larger values of N_P, so that the expansion can be truncated to the finite dimension N_P. Usually, such a series representation requires additional regularity of X, e.g., X should be square-integrable with respect to the underlying probability space in order to preserve the convergence in $\mathcal{L}_2$ (see the convergence of RV in appendix A).

The exploitation of the *Stochastic Spectral Method* (SSM) can be viewed from two perspectives. The first perspective is that the SSM is used for assessing the pdf of QoI in uncertainty propagation. The global sensitivity measure, the *Sobol Index*, can be also determined directly from the spectral representation. This can be achieved by exploiting the smoothness usually exhibited by a high-dimensional parameter space to construct a solution with convergences rates significantly better than with primitive sampling based method. The other perspective is that the SSM can be used to construct a surrogate model. In the probabilistic UQ approaches, the computation of the model is expanded to a stochastic parameter space. This usually requires more computational efforts compared to the computation

1 For clarification, in this setting the coefficients x_k itself is deterministic.

in deterministic cases. Using the surrogate model constructed by SSM facilitates the procedures such as Bayesian model calibration, sensitivity analysis design and control implementation.

By using the Hermite polynomials as the basis functions $\Xi_k(\cdot)$, these expansions are termed as *Polynomial Chaos Expansion* (PCE). Ghanem and Spanos applied the original PCE to quantify the uncertainty in solid mechanics system by applying the PCE to the Finite-Element discretization [Gha03]. Xiu and Karniadakis extended the PCE to some parametric non-Gaussian random processes [Xiu02] and named it as *generalized Polynomial Chaos* (gPC). In case that other basis functions are used, e.g. Wavelet [LM10], Padè [Cha09], Fourier (sines and cosines) etc., these expansions are also covered by the generic term as Stochastic Spectral Method (SSM). As the probability distributions established in gPCE is related to the pdfs used in the developed framework , the PCE and its generalized variant are mainly discussed in this thesis. The mathematical fundamentals of gPC described in following are summarized from [Smi13], [LM10] and [Xiu10].

3.1.1 Polynomial Chaos and generalized Polynomial Chaos

Consider an arbitrary real-valued random variable $X = X(\omega)$ according to some probability space $(\Omega,\mathfrak{A},\mathcal{P})$, with sample space Ω, σ-algebra $\mathfrak{A}$, and probability measure $\mathcal{P}$ (see appendix A for more detailed definition). For the stochastic formulation discussed here, it is required that the RVs are square-integrable and have a finite variance, i.e., $X \in \mathcal{L}_2(\Omega,\mathfrak{A},\mathcal{P}) = \{X : \mathbb{E}(X^2) < \infty\}$.

Let $\{\xi_i\}_{i=1}^{\infty}$ be a sequence of centered, normalized, mutually orthogonal Gaussian variables. Let $\hat{\mathbb{P}}_P$ denote the space of polynomials in $\{\xi_i\}_{i=1}^{\infty}$ having degree less or equal to $P \in \mathbb{N}$. Furthermore, let $\Xi_P \subset \hat{\mathbb{P}}_P$ be the set of polynomials that belong to $\hat{\mathbb{P}}_P$ and are orthogonal to $\hat{\mathbb{P}}_{P-1}$, and define $\breve{\mathbb{P}}_P$ as the space spanned by Ξ_P. It yields:

$$\hat{\mathbb{P}}_P = \hat{\mathbb{P}}_{N-1} \oplus \breve{\mathbb{P}}_P, \qquad \mathcal{L}_2(\Omega,\mathfrak{A},\mathcal{P}) = \bigoplus_{i=0}^{\infty} \breve{\mathbb{P}}_i \tag{3.2}$$

The subspace $\breve{\mathbb{P}}_P$ of $\mathcal{L}_2(\Omega,\mathfrak{A},\mathcal{P})$ is called the P-th Homogeneous Chaos, and Ξ_P is called the *Polynomial Chaos* (PC) of order P. The PC expansion of a random variable X is:

$$X(\omega) = a_0\Xi_0 + \sum_{i_1=1}^{\infty} a_{i_1}\Xi_1\left(\xi_{i_1}(\omega)\right) + \sum_{i_1=1}^{\infty}\sum_{i_2=1}^{i_1} a_{i_1 i_2}\Xi_2\left(\xi_{i_1}(\omega),\xi_{i_2}(\omega)\right)$$
$$+\sum_{i_1=1}^{\infty}\sum_{i_2=1}^{i_1}\sum_{i_3=1}^{i_2} a_{i_1 i_2 i_3}\Xi_3\left(\xi_{i_1}(\omega),\xi_{i_2}(\omega),\xi_{i_3}(\omega)\right) + \ldots \quad (3.3)$$

The Polynomial Chaos introduced by Wiener [Wie38] are constructed from the *Hermite Polynomial* $H_k(\cdot)$. Cameron and Martin proved that the expression (3.3), in case of the PC constructed from the Hermite Polynomial, is convergent for any RVs $X(\omega)$ in the Hilbert space $\mathcal{L}_2(\Omega)$ [Cam47].

Generalized Polynomial Chaos

Being a spectral polynomial expansion this expansion has an exponential convergent rate. However, for non-Gaussian random variables, the expansion may exhibit low convergence rates and thus require a high truncation order. Therefore, Xiu and Karniadakis [Xiu02] employed the Askey-scheme to generalize the original Wiener's PC expansion to some common non-Gaussian measures, which replaces the Hermite polynomial by other orthogonal polynomials and named it as *generalized Polynomial Chaos Expansion* (gPCE). Xiu concluded that the Cameron-Martin theorem (Eq. (3.3)) can be generalized to the gPCE, because each type of polynomials from the Askey scheme form a complete basis in the Hilbert space determined by their corresponding random vector $\boldsymbol{\xi}$ [Xiu04a]. Table 3.1 shows the correspondences between the random variable distribution and orthogonal polynomial family.

The distributions established in the gPC expansion match the distribution used in the developed framework in this thesis well. As discussed in the section 2.1.5, the probability distribution functions, which can be derived from the typical information found in engineering are (a) the normal distribution, (b) the uniform distribution and (c) the exponential distribution. The exponential distribution is a special case of the gamma distribution.

	pdf of RV $\xi(\omega)$	gPC basis polynomials $\psi(\xi)$	Support
Continuous	Gaussian	Hermite	$(-\infty,\infty)$
	Gamma	Laguerre	$[0,\infty)$
	Beta	Jacobi	$[a,b]$
	Uniform	Legendre	$[a,b]$
Discrete	Poisson	Charlier	$\{0,1,2,\dots\}$
	Binomial	Krawtchouk	$\{0,1,2,\dots,n\}$
	Negative Binomial	Meixner	$\{0,1,2,\dots\}$
	Hypergeometric	Hahn	$\{0,1,2,\dots,n\}$

Table 3.1: Correspondence between the type of generalized Polynomial Chaos and their underlying random variables [Xiu10]

Single random variable

The RV $X(\omega)$ can be expanded by one-dimensional gPC basis functions as:

$$X(\omega) = \sum_{k=0}^{\infty} a_k \psi_k(\xi(\omega)) \tag{3.4}$$

where $\psi_k(\cdot)$ denotes the one-dimensional orthogonal polynomial with the degree of k, e.g. Hermite Polynomials $H_k(\cdot)$, Legendre Polynomials $Le_k(\cdot)$, or Lagauerre Polynomials $La_k(\cdot)$.

Multivariate random variable

Now the one-dimensional gPC basis is extended to the d-dimensional PC-basis. Given $\boldsymbol{\xi}(\boldsymbol{\omega}) = (\xi_1, \xi_2, \dots, \xi_d)$ a set of centered, normalized and mutually orthogonal Gaussian random variables, the PC expansion (3.3) has a form:

$$\begin{aligned} X(\omega) = a_0 \Xi_0 &+ \sum_{i_1=1}^{d} a_{i_1} \Xi_1\left(\xi_{i_1}(\omega)\right) + \sum_{i_1=1}^{d} \sum_{i_2=1}^{i_1} a_{i_1 i_2} \Xi_2\left(\xi_{i_1}(\omega), \xi_{i_2}(\omega)\right) \\ &+ \sum_{i_1=1}^{d} \sum_{i_2=1}^{i_1} \sum_{i_3=1}^{i_2} a_{i_1 i_2 i_3} \Xi_3\left(\xi_{i_1}(\omega), \xi_{i_2}(\omega), \xi_{i_3}(\omega)\right) + \dots \end{aligned} \tag{3.5}$$

To simplify the notation, the multi-index notation $\underline{i} = (i_1, i_2, \ldots, i_\alpha, \ldots, i_d)$ with $|\underline{i}| = \sum_{\alpha=1}^{d} i_\alpha$ is adopt. Let $\underline{i}(\alpha)$ denote the α-th element of the multi-index $\underline{i}$. For construction gPC, the set of multi-indices $\varrho(P)$ is defined as:

$$\varrho(P) := \{\underline{i} : |\underline{i}| = P\}. \tag{3.6}$$

The P-th Polynomial Chaos Ξ_P in equation (3.5) is constructed from one-dimensional gPC basis functions as:

$$\Xi_P := \left\{ \bigcup_{\underline{i} \in \varrho(P)} \prod_{\alpha=1}^{d} \psi_{\underline{i}(\alpha)}(\xi_\alpha) \right\}. \tag{3.7}$$

The expansion (3.5) can be rewritten in more compact form as

$$X(\omega) = \sum_{|\underline{i}|=0}^{\infty} \beta_{\underline{i}} \Psi_{\underline{i}}(\xi_1, \xi_2, \ldots, \xi_d), \tag{3.8}$$

where $\beta_{\underline{i}}$ are the deterministic expansion coefficients and $\Psi_{\underline{i}}$ are polynomials constructed by tensor products of the 1D-polynomials $\psi_{\underline{i}(\alpha)}(\xi_\alpha)$

$$\Psi_{\underline{i}}(\xi_1, \xi_2, \ldots, \xi_d) = \prod_{\alpha=1}^{d} \psi_{\underline{i}(\alpha)}(\xi_\alpha). \tag{3.9}$$

Although using multi-index formulation is very clear, the single index is preferable to express the gPC expansion. The multi-index can be converted to the single index. To do this, the lexicographic order is applied as also used by [Xiu10]. The relation between the single index k, the multi-index $\underline{i}$ and the orthogonal polynomial Ψ_k is established in table 3.2. By using single index k, equation (3.8) is rewritten as

$$X(\omega) = \sum_{k=0}^{\infty} \beta_k \Psi_k(\xi_1, \xi_2, \ldots, \xi_d). \tag{3.10}$$

With the spectral convergence property of the orthogonal projection, the expansion can be truncated to finite dimension. The finite-dimensional decomposition of $X(\omega)$ in the single-index form is:

$$X(\omega) \approx X_P(\omega) = \sum_{k=0}^{N_P} \beta_k \Psi_k(\boldsymbol{\xi}(\omega)), \tag{3.11}$$

where the basis dimension N_P is related to the dimension of the multivariate random variable d and the polynomial order P by

$$N_P + 1 = \frac{(P+d)!}{P!d!}. \tag{3.12}$$

$\lvert \underline{i} \rvert$	$\underline{i} = (i_1, i_2, \ldots, i_d)$	**Polynomial** Ψ_k	k
0	(0,0,...,0,0)	$\psi_0(\xi_1)\psi_0(\xi_2)\cdots\psi_0(\xi_d) = 1$	0
1	(1,0,...,0,0)	$\psi_1(\xi_1)\psi_0(\xi_2)\cdots\psi_0(\xi_d) = \psi_1(\xi_1)$	1
	(0,1,...,0,0)	$\psi_0(\xi_1)\psi_1(\xi_2)\cdots\psi_0(\xi_d) = \psi_1(\xi_2)$	2
	$\vdots$	$\vdots$	$\vdots$
	(0,0,...,0,1)	$\psi_0(\xi_1)\psi_0(\xi_2)\cdots\psi_1(\xi_d) = \psi_1(\xi_d)$	$\binom{d+1}{d} - 1 = d$
2	(2,0,0,0,...,0)	$\psi_2(\xi_1)\psi_0(\xi_2)\psi_0(\xi_3)\psi_0(\xi_4)\cdots\psi_0(\xi_d)$	$d+1$
	(1,1,0,0,...,0)	$\psi_1(\xi_1)\psi_1(\xi_2)\psi_0(\xi_3)\psi_0(\xi_4)\cdots\psi_0(\xi_d)$	$d+2$
	(1,0,1,0,...,0)	$\psi_1(\xi_1)\psi_0(\xi_2)\psi_1(\xi_3)\psi_0(\xi_4)\cdots\psi_0(\xi_d)$	
	$\vdots$	$\vdots$	
	(1,0,0,...,0,1)	$\psi_1(\xi_1)\psi_0(\xi_2)\psi_0(\xi_3)\cdots\psi_0(\xi_{d-1})\psi_1(\xi_d)$	$\vdots$
	$\vdots$	$\vdots$	
	(0,0,0,...,0,2)	$\psi_0(\xi_1)\psi_0(\xi_2)\psi_0(\xi_3)\cdots\psi_0(\xi_{d-1})\psi_2(\xi_d)$	$\binom{d+2}{d} - 1$
3	(3,0,...,0)	$\psi_3(\xi_1)\psi_0(\xi_2)\psi_0(\xi_3)\psi_0(\xi_4)\cdots\psi_0(\xi_d)$	$\binom{d+2}{d}$
	$\vdots$	$\vdots$	$\vdots$

Table 3.2: Single index, multi-index, and tensored polynomials

The polynomials $\Psi_k(\cdot)$ form an orthogonal and complete basis in $\mathcal{L}_2$. With respect to the inner product, the orthogonality is expressed as:

$$\begin{aligned} \langle \Psi_i | \Psi_j \rangle \equiv \int \Psi_j(\boldsymbol{\xi})\Psi_i(\boldsymbol{\xi}) p(\boldsymbol{\xi}) d\boldsymbol{\xi} &= \delta_{ij} \langle \Psi_i | \Psi_i \rangle \\ &= \delta_{ij} \left\| \Psi_i^2 \right\| = \gamma_i \end{aligned} \tag{3.13}$$

Because of the orthogonal property, the statistical values can be calculated directly from the PCE coefficients. The mean of the RV $X(\omega)$ is given by

$$\begin{aligned} \overline{X} = \mathbb{E}(X(\omega)) &= \mathbb{E}\left[\sum_{k=0}^{\infty} \beta_k \Psi_k(\boldsymbol{\xi}(\omega)) \right] \\ &= \sum_{k=0}^{\infty} \beta_k \mathbb{E}\left[\Psi_k(\boldsymbol{\xi}(\omega)) \cdot 1 \right] \end{aligned}$$

$$= \sum_{k=0}^{\infty} \beta_k \langle \Psi_k | \Psi_0 \rangle$$

$$\overline{X} = \beta_0. \tag{3.14}$$

Similarly the variance of the RV $\sigma_X^2 = \mathrm{Var}[X(\omega)]$ is:

$$\begin{aligned}
\mathrm{Var}[X(\omega)] &= \mathbb{E}\left[\left(X(\omega) - \overline{X}\right)^2\right] \\
&= \mathbb{E}\left[\left(\sum_{k=0}^{\infty} \beta_k \Psi_k(\boldsymbol{\xi}(\omega)) - \beta_0\right)^2\right] \\
&= \mathbb{E}\left[\left(\sum_{k=1}^{\infty} \beta_k \Psi_k(\boldsymbol{\xi}(\omega))\right)^2\right] \\
&= \sum_{k=1}^{\infty} \sum_{l=1}^{\infty} \beta_k \beta_l \mathbb{E}\left[\Psi_k(\boldsymbol{\xi}(\omega)) \Psi_l(\boldsymbol{\xi}(\omega))\right] \\
&= \sum_{k=1}^{\infty} \sum_{l=1}^{\infty} \beta_k \beta_l \langle \Psi_k | \Psi_l \rangle \\
&= \sum_{k=1}^{\infty} \beta_k^2 \|\Psi_k\|^2, \\
\sigma_X^2 &\approx \sum_{k=1}^{N_P} \beta_k^2 \|\Psi_k\|^2.
\end{aligned} \tag{3.15}$$

Dependent random variables

The PCE requires the mutually independent random variables and the representation of the joint density $p_{\boldsymbol{\xi}}(\boldsymbol{\xi}) = \prod_{i=1}^{d} p_{\xi_i}(\xi_i)$ for implementation as assumed in the course of this thesis. For correlated RV ξ_i, the *Nataf* transformation in a combination with a *Cholesky* decomposition can be employed to construct mutually independent Gaussian random variables, if marginal distributions and a correlation matrix are available. The reader is referred to [LM10] for more details about the PCE of dependent RV.

3.1.2 Spectral representation of a random vector

The gPC expansion of a random variable (eq. (3.10)) can be extended to represent $\mathbb{R}^n$-random vectors $\boldsymbol{X} : \Omega \mapsto \mathbb{R}^n$ (see Def. 5). In the same manner as in eq. (3.11), the truncated gPCE of the i-th component of the random vector X_i is

$$X_i \approx \sum_{k=0}^{N_P} \beta_{ik} \Psi_k(\boldsymbol{\xi}).$$

The random vector can be expanded to:

$$\boldsymbol{X}(\omega) \approx \sum_{k=0}^{N_P} \boldsymbol{\beta}_k \Psi_k(\boldsymbol{\xi}). \tag{3.16}$$

where $\boldsymbol{\beta}_k = \left[\beta_{1k} \cdots \beta_{nk}\right]^T \in \mathbb{R}^n$ consists of the coefficients of the k-th PC basis of the random vector components. The vector $\boldsymbol{\beta}_k$ is called the k-th stochastic mode of the random vector $\boldsymbol{X}$. Regarding eq. (3.14), $\boldsymbol{\beta}_0$ is the mean of the random vector.

In addition, components X_i and X_j are orthogonal if and only if

$$\sum_{k=0}^{N_P} \beta_{ik} \beta_{jk} \langle \Psi_k^2(\xi) \rangle = 0.$$

The covariance matrix of the random vector $\boldsymbol{X}$ is given by (see Def. 11):

$$\mathbf{C}_{XX} = \mathbb{E}\left[\tilde{\boldsymbol{X}} \otimes \tilde{\boldsymbol{X}}\right]. \tag{3.17}$$

where $\tilde{\boldsymbol{X}} = \boldsymbol{X} - \mathbb{E}(\boldsymbol{X})$ is defined as the fluctuations of the random vector $\boldsymbol{X}$. If $\boldsymbol{X}$ is represented in the spectral form as in eq. (3.16), it yields the relation:

$$\tilde{\boldsymbol{X}} = \boldsymbol{X} - \mathbb{E}(\boldsymbol{X}) \approx \sum_{k=1}^{N_P} \boldsymbol{\beta}_k \Psi_k(\boldsymbol{\xi}). \tag{3.18}$$

As mentioned in [Paj12], the PCE representation (3.16) leads to a covariance matrix of the form:

$$\begin{aligned}\mathbf{C}_{XX} &= \mathbb{E}\left[\left(\sum_{k>0}\boldsymbol{\beta}_k\Psi_k(\boldsymbol{\xi})\right)\otimes\left(\sum_{l>0}\boldsymbol{\beta}_l\Psi_l(\boldsymbol{\xi})\right)\right]\\ &= \sum_{k>0}\sum_{l>0}\left((\boldsymbol{\beta}_k\otimes\boldsymbol{\beta}_l)\mathbb{E}\left[\Psi_k(\boldsymbol{\xi}(\omega))\,\Psi_l(\boldsymbol{\xi}(\omega))\right]\right)\\ &\approx \sum_{k>0}^{P}\sum_{l>0}^{P}\left((\boldsymbol{\beta}_k\otimes\boldsymbol{\beta}_l)\mathbb{E}\left[\Psi_k(\boldsymbol{\xi}(\omega))\,\Psi_l(\boldsymbol{\xi}(\omega))\right]\right).\end{aligned} \tag{3.19}$$

The diagonal Gram $(N_P \times N_P)$ matrix $\mathbf{G} = (g_{kl})$ is introduced with $g_{kl} = \mathbb{E}[\Psi_k(\omega)\Psi_l(\omega)]$. The equation (3.19) can be rewritten in matrix form as:

$$\mathbf{C}_{XX} = \check{\mathbf{X}}\mathbf{G}\check{\mathbf{X}}^T, \tag{3.20}$$

where $\check{\mathbf{X}} = [\boldsymbol{\beta}_1, \ldots, \boldsymbol{\beta}_{N_P}]$ is a $(n \times N_P)$ matrix containing PC coefficients of the random vector without the $\boldsymbol{\beta}_0$ term, i.e. without the mean $\overline{\boldsymbol{X}}$.

3.1.3 PC expansion of random processes

The random process $\mathcal{X}(\boldsymbol{r},\omega)$ in $\mathcal{L}_2$ can be separated into the deterministic and the stochastic parts with the stochastic discretization methods mentioned in the section 2.2.3. The polynomial chaos expansion is mentioned in the section as one of the **series expansion methods**. The PC expansion of the random process $\mathcal{X}(\boldsymbol{r},\omega)$ can be considered as the generalization of the PC expansion of the random vector, namely

$$\mathcal{X}(\boldsymbol{r},\omega) = \sum_{k=0}^{\infty} x_k(\boldsymbol{r})\Psi_k(\boldsymbol{\xi}(\omega)) \approx \sum_{k=0}^{N_P} x_k(\boldsymbol{r})\Psi_k(\boldsymbol{\xi}(\omega)), \tag{3.21}$$

where the deterministic function $x_k(\boldsymbol{r})$ are called the stochastic modes of the process. It is noted that the PC expansion contains more information than the second-order properties of random field $\mathcal{X}(\boldsymbol{r},\omega)$. The knowledge of the correlation function is not sufficient to uniquely determine the set of coefficients $x_k(\boldsymbol{r})$ in the expansion 3.21.

For computational purposes, the stochastic modes are commonly discretized in the spatial domain with some basis function $\phi_i(\boldsymbol{r})$ (cf. eq. (2.11)).

This leads equation (3.21) to

$$\mathcal{X}(\boldsymbol{r},\omega) = \sum_{k=0}^{N_P} \overbrace{\left(\mathbf{x}_k \cdot \boldsymbol{\phi}(\boldsymbol{r})\right)}^{x_k(\boldsymbol{r})} \Psi_k(\boldsymbol{\xi}(\omega)), \tag{3.22}$$

where $\boldsymbol{\phi}(\boldsymbol{r}) = \left[\phi_1(\boldsymbol{r}),\ldots,\phi_{N_s}(\boldsymbol{r})\right]^T$ denotes the vector of shape functions and $\mathbf{x}_k = \left[x_{k1},\ldots,x_{kN_s}\right]^T$ represents corresponding coefficients. Equation (3.21) can be rewritten as:

$$\mathcal{X}(\boldsymbol{r},\omega) = \underbrace{\sum_{k=0}^{N_P} \left(\mathbf{x}_k \Psi_k(\boldsymbol{\xi}(\omega))\right)}_{\text{PCE of random vector}} \cdot \boldsymbol{\phi}(\boldsymbol{r}). \tag{3.23}$$

Equation (3.23) shows the relation to the PC expansion of the random vector (3.16). This can be seen as the PC expansion with the method of point discretization. This approach is simple to compute and not restricted to Gaussian random fields. However, the method is only useful for medium to long correlation distances. One may need a fine mesh to fulfill the accuracy requirements in case of a small correlation length. A large number of discretization points N_s requires large computational effort.

As discussed in section 2.2.3, the KLE offers an efficient expansion. The KLE can be combined with the PC expansion [Ros12a]. The combination of the KLE and the PC expansion is discussed following.

KLE/PC expansion

Using the KLE method to discretize a random field has been discussed in section 2.2.3. The random field can be approximated in the form:

$$\mathcal{X}(\boldsymbol{r},\omega) \approx \sum_{i=0}^{N_{KLE}} \sqrt{\lambda_i}\theta_i(\omega)\phi_i(\boldsymbol{r}). \tag{3.24}$$

The PC expansion can be extended to stochastic processes by approximating the RV $\theta_i(\omega)$ by a convergent polynomial chaos expansion:

$$\theta_i(\omega) \approx \sum_{k=0}^{N_P} \beta_k^{(i)} \Psi_k(\boldsymbol{\xi}(\omega)). \tag{3.25}$$

Substituting the PCE approximation in the Eq. (3.24) yields:

$$\mathcal{X}(\boldsymbol{r},\omega) \approx \sum_{i=1}^{N_{KLE}} \sqrt{\lambda_i} \left(\sum_{k=0}^{N_P} \beta_k^{(i)} \Psi_k(\boldsymbol{\xi}(\omega)) \right) \phi_i(\boldsymbol{r}). \tag{3.26}$$

Equation (3.26) can be rearranged in the form:

$$\mathcal{X}(\boldsymbol{r},\omega) \approx \sum_{k=0}^{N_P} \left(\sum_{i=1}^{N_{KLE}} \sqrt{\lambda_i} \beta_k^{(i)} \phi_i(\boldsymbol{r}) \right) \Psi_k(\boldsymbol{\xi}(\omega)), \tag{3.27}$$

which is equivalent to equation (3.21) in case of

$$x_k(\boldsymbol{r}) = \left(\sum_{i=1}^{N_{KLE}} \sqrt{\lambda_i} \beta_k^{(i)} \phi_i(\boldsymbol{r}) \right). \tag{3.28}$$

3.2 Uncertainty quantification with gPC

Considering the state $x(\boldsymbol{r},t)$ of the DPS that is the solution of the equation

$$\mathcal{M}_s\left(x(\boldsymbol{r},t),\boldsymbol{q}\right) = 0$$

as described in the section 2.1.3. In the Bayesian framework of analysis and identification, the d-dimensional parameter vector $\boldsymbol{q}$ is considered to be uncertain and is described with the random vector $\boldsymbol{Q}(\omega) = [Q_1(\omega),\ldots,Q_d(\omega)]^T$. The solution of the PDE is a functional of the random vector $\boldsymbol{Q}(\omega)$ and can be considered as a random process $\mathcal{X}(\boldsymbol{r},t,\omega)$.

As discussed in section 2.1.5, the pdf $p_{\boldsymbol{Q}}(\boldsymbol{q})$ is derived from the available knowledge by using the Principle of Maximum Entropy. For the uniform, normal and exponential distributions, which can be found in many applications in engineering, the gPCE is an attractive method. The gPCE approximates the random process $\mathcal{X}(\boldsymbol{r},t,\omega)$ in terms of the series of the orthogonal polynomials according to the Askey-scheme with corresponding coefficients.

$$\mathcal{X}(\boldsymbol{r},t,\omega) \approx \mathcal{X}_{N_P}(\boldsymbol{r},t,\omega) = \sum_{k=0}^{N_P} \beta_k(\boldsymbol{r},t)\Psi_k(\boldsymbol{\xi}(\omega))$$

It is noted that for computational purposes the gPC coefficients $\beta_k(\boldsymbol{r},t)$ are discretized in spatial and temporal space as described in the sections 2.2.1, 2.2.2 and recasted in form of vectors.

3.2.1 Computation of gPC coeficients

Since the gPC coefficients β_k characterize the random variable $X(\omega)$, a procedure for the determination of these gPC coefficients is needed. In the UQ community, three important approaches are proposed for the determination of the gPC coefficients, namely *Galerkin projection, stochastic collocation* and *pseudo discrete projection.* Regarding to their implementations, these three approaches are classified as *intrusive* and *non-intrusive* methods. A method is considered as intrusive, if the existing code of a deterministic solver has to be modified to obtain the stochastic solution. On the contrary, non-intrusive methods need only solutions of the inputs realizations to determine the gPC coefficients.

The Galerkin projection is considered as an intrusive method, whereas the stochastic collocation and the pseudo discrete projection belong to the class of non-intrusive methods. All three approaches and their attributes are discussed in this subsection.

Stochastic Galerkin method

Substituting the truncated PCE expression X_P to X in equation (2.4), the model equation is not satisfied anymore but yields a residual. By the Galerkin projection, the residual has to be orthogonal to the space of the expansion basis functions $\{\Psi_k(\boldsymbol{q})\}_{k=0}^{N_P}$ (cf. Eq. (2.14)). It yields

$$\left\langle \mathcal{M}_s\left(\sum_{k=0}^{N_P} \beta_k \Psi_k(\boldsymbol{q}), \boldsymbol{q}\right) \middle| \Psi_k \right\rangle = 0, \qquad k = 0, \ldots, N_P. \tag{3.29}$$

The Galerkin method leads to a set of $N_P + 1$ coupled problems. In order to solve Eq. (3.29), it usually requires modifications of the existing numerical code. If the code is complex, the Galerkin procedure can be difficult to implement and therefore is not practical in general. In practice the non-intrusive methods are applied more, as one wants to avoid the modification of the existing deterministic code.

Stochastic collocation method

The gPC coefficients can be estimated by using a regression method. Denoting $\{\boldsymbol{q}^r\}_{r=1}^{N_r}$ a sample set of the RVs, also called collocation points, and $\{x^r\}_{r=1}^{N_r}$ the corresponding set of quantities of interest (QoI), so that

$$\mathcal{M}_s\left(x^r, \boldsymbol{q}^r\right) = 0, \ \forall r. \tag{3.30}$$

The sought PC coefficients in the truncated expansion of X are rewritten in vector form $\boldsymbol{\beta} = (\beta_0, \ldots, \beta_{N_P})^T$. Based on the sample set $\{\boldsymbol{q}^r\}_{r=1}^{N_r}$ the optimal approximation $\hat{\boldsymbol{\beta}}$ of $\boldsymbol{\beta}$ can be obtained by solving the least square problem

$$\hat{\boldsymbol{\beta}} = \arg\min_{\boldsymbol{\beta}} \sum_{r=1}^{N_r} \left(x^r - \sum_{k=0}^{N_P} \beta_k \Psi_k(\boldsymbol{q}^r) \right)^2 . \tag{3.31}$$

The corresponding QoIs x^r to the samples are arranged in the vector form $\mathbf{x} = [x^1 \ \cdots \ x^{N_r}]^T$, so that the solution of the least squares problem (3.31) is

$$\hat{\boldsymbol{\beta}} = \left(\mathbf{Z}^T\mathbf{Z}\right)^{-1} \mathbf{Z}^T \mathbf{x}, \tag{3.32}$$

where

$$\mathbf{Z} = \begin{pmatrix} \Psi_0(\boldsymbol{q}^1) & \Psi_1(\boldsymbol{q}^1) & \cdots & \Psi_{N_P}(\boldsymbol{q}^1) \\ \Psi_0(\boldsymbol{q}^2) & \Psi_1(\boldsymbol{q}^2) & \cdots & \Psi_{N_P}(\boldsymbol{q}^2) \\ \vdots & \vdots & \ddots & \vdots \\ \Psi_0(\boldsymbol{q}^{N_r}) & \Psi_1(\boldsymbol{q}^{N_r}) & \cdots & \Psi_{N_P}(\boldsymbol{q}^{N_r}) \end{pmatrix}. \tag{3.33}$$

The method depends on the sample sizes, due to the fact that the approximation space changes as the number of collocation points changes [Smi13]. For the d-dimensional multivariate random variable and P-dimensional basis, an empirical rule for the optimal number of regression points N_r is given by $N_r = N_P \cdot (d-1)$ [Sud08]. The sample set can be constructed by *Simple Random Sampling* (SRS), *Latin Hypercube Sampling* (LHS), *Quasi-Monte-Carlo* (QMC) etc. The methods from statistical learning, e.g. *adjusted* R^2, *early stopping* etc., can be applied to avoid overfitting [Cre09, Lew07].

Pseudo discrete projection

In contrast to the projection the residual to governing equation as Galerkin projection, the *Pseudo discrete projection,* also known as the *non-intrusive spectral projection* (NISP) exploits the orthogonality of the gPC basis by projecting directly onto the QoI, by taking the inner product of the output PC expansion with orthogonal polynomial Ψ_k

$$\langle x(\boldsymbol{q})|\Psi_k\rangle = \left\langle \sum_{k=0}^{N_P} \beta_k \Psi_k(\boldsymbol{q}) \middle| \Psi_k(\boldsymbol{q}) \right\rangle, \tag{3.34}$$

where the definition of the inner product is

$$\langle f(\boldsymbol{q})|g(\boldsymbol{q})\rangle = \int_{\Omega^d} f(\boldsymbol{q}) g(\boldsymbol{q}) p_Q(\boldsymbol{q}) d\boldsymbol{q}. \tag{3.35}$$

Using the orthogonality property, it yields

$$\beta_k = \frac{\langle x(\boldsymbol{q})|\Psi_k(\boldsymbol{q})\rangle}{\langle \Psi_k(\boldsymbol{q})|\Psi_k(\boldsymbol{q})\rangle} = \frac{\langle x(\boldsymbol{q})|\Psi_k(\boldsymbol{q})\rangle}{\|\Psi_k(\boldsymbol{q})\|^2}. \tag{3.36}$$

Thanks to the polynomial property of the Ψ_k, the term $\|\Psi_k(\boldsymbol{q})\|^2 = \gamma_k$ (see Appendix C) can be evaluated analytically. The determination of the PC coefficients requires also the evaluation of $d-$dimensional integrals:

$$\langle x(\boldsymbol{q})|\Psi_k(\boldsymbol{q})\rangle = \int_{\Omega^d} x(\boldsymbol{q}) \Psi_k(\boldsymbol{q}) p_Q(\boldsymbol{q}) d\boldsymbol{q} \tag{3.37}$$

for $k = 0 \ldots N_P$. The integration is accomplished mostly by numerical integration methods, which are discussed in section 3.2.4.

3.2.2 gPC expansion of convection-diffusion systems

In this section, the gPC expansion of the convection-diffusion system as described in section 2.1.3 is demonstrated. All the parameters in the IBVP are collected in the parameter vector $\boldsymbol{q}$, which is uncertain regarding the developed framework and described with the random vector $\boldsymbol{Q}(\omega)$. Therefore, equations (2.6), (2.2) and (2.3) can be written in form:

$$\frac{\partial x}{\partial t} = \nabla \cdot (v(\boldsymbol{Q})\nabla x) - \nabla \cdot (\boldsymbol{v}(\boldsymbol{Q})x) + s(\boldsymbol{Q}), \qquad \boldsymbol{r} \in \mathcal{G}, t \in \mathcal{T} \tag{3.38}$$

$$\mathcal{B}(x(\boldsymbol{r},t),\boldsymbol{Q}) = b(t,\boldsymbol{Q}), \qquad \boldsymbol{r} \in \partial\mathcal{G}, t \in \mathcal{T} \tag{3.39}$$

$$x(\boldsymbol{r},t=0) = g(\boldsymbol{r},\boldsymbol{Q}), \qquad \boldsymbol{r} \in \mathcal{G} \tag{3.40}$$

The solution of the IBVP depends on the random vector $\boldsymbol{Q}(\omega)$, so the solution $x(\boldsymbol{r},t,\boldsymbol{Q})$ can be considered as a stochastic process. In order to discretize $x(\boldsymbol{r},t,\boldsymbol{Q})$, the finite-dimensional subspace $V^{N_s} = \operatorname{span}\{\phi_j\}_{j=1}^{N_s}$ and $\breve{\mathbb{P}}_P = \operatorname{span}\{\Psi_k\}_{k=1}^{N_P}$ are constructed, where ϕ_j are spatial basis functions and orthogonal polynomials $\Psi_k(\boldsymbol{Q})$ are used as a basis for the random components. According to equation (3.21), the approximation solution is:

$$\hat{x}(t,\boldsymbol{r},\boldsymbol{Q}) = \sum_{k=0}^{N_P} \sum_{i=1}^{N_s} \beta_{ik}(t)\phi_i(\boldsymbol{r})\Psi_k(\boldsymbol{Q}). \tag{3.41}$$

The gPC coefficients $\beta_{ik}(t)$, which characterize the solution, can be computed by the three methods mentioned in the preceding section. The formulations of each method to the convection-diffusion system are demonstrated in following.

Stochastic Galerkin

Substituting the approximation eq. (3.41) in eq. (3.38) and applying the basis functions as the test function results:

$$\int_{\Omega} \Psi_k(\boldsymbol{Q}) \int_{\mathcal{G}} \phi_i(\boldsymbol{r}) \Big(\frac{\partial}{\partial t}\hat{x}(\boldsymbol{r},t,\boldsymbol{Q}) - \nabla \cdot (v(\boldsymbol{Q})\nabla \hat{x}(\boldsymbol{r},t,\boldsymbol{Q}))$$
$$+\nabla \cdot (\boldsymbol{v}(\boldsymbol{Q})\hat{x}(\boldsymbol{r},t,\boldsymbol{Q})) - s(\boldsymbol{r},t,\boldsymbol{Q}) \Big) d\boldsymbol{r}\, p_{\boldsymbol{Q}}(\boldsymbol{q}) d\boldsymbol{q} = 0$$
$$\text{for } i = 1,\ldots,N_s \text{ and } k = 0,\ldots,N_P \tag{3.42}$$

As shown in section 2.2.1, the integration in the spatial domain can be rewritten in a matrix form. In this case the matrix depends on the random vector $\boldsymbol{Q}$, it yields:

$$\int_{\Omega} \Psi_k(\boldsymbol{Q})\Bigg(\mathbf{M}(\boldsymbol{Q})\frac{\partial}{\partial t}\left(\sum_{k=0}^{N_P} \boldsymbol{\beta}_k(t)\Psi_k(\boldsymbol{Q})\right)+$$

$$(\mathbf{N}(\boldsymbol{Q})+\mathbf{G}(\boldsymbol{Q}))\left(\sum_{k=0}^{P} \boldsymbol{\beta}_k(t)\Psi_k(\boldsymbol{Q})\right)\Bigg)p_{\boldsymbol{Q}}(\boldsymbol{q})d\boldsymbol{q} = \int_{\Omega} \Psi_k(\boldsymbol{Q})\mathbf{b}(\boldsymbol{Q})p_{\boldsymbol{Q}}(\boldsymbol{q})d\boldsymbol{q}, \tag{3.43}$$

where $\boldsymbol{\beta}_k(t) = \left[\beta_{1k}(t),\ldots,\beta_{N_s k}(t)\right]^T$ contains the corresponding coefficients. As discussed in 2.2.1, the Galerkin projection for a deterministic case results in a system of N_s ordinary differential equations. The stochastic Galerkin projection eq. (3.43) constructs a system of ODEs with $(N_s \times (N_P+1))$ coupled equations to solve for the coefficients $x_{ik}(t)$.

Stochastic collocation

Let $\{\boldsymbol{q}^r\}_{r=1}^{N_r}$ be a sample set of the random variables from the pdf $p_{\boldsymbol{Q}}(\boldsymbol{q})$, and $\mathbf{x}(t,\boldsymbol{q}^r)$ is the corresponding state, which is the solution of the equation

$$\mathbf{M}(\boldsymbol{q}^r)\frac{\partial}{\partial t}\mathbf{x}(t,\boldsymbol{q}^r) + \left(\mathbf{N}(\boldsymbol{q}^r)+\mathbf{G}(\boldsymbol{q}^r)\right)\mathbf{x}(t,\boldsymbol{q}^r) = \mathbf{b}(\boldsymbol{q}^r). \tag{3.44}$$

In the time dependent problem, the solution $\mathbf{x}(t)$ is represented by the series $\mathbf{x}(t_j)$ where $t_j = j\Delta t$. By defining the block-matrix $\mathbf{X}$ as

$$\mathbf{X} = \begin{bmatrix} x_1(t_1,\boldsymbol{q}^1) & \cdots & x_1(t_1,\boldsymbol{q}^r) & \cdots & x_1(t_1,\boldsymbol{q}^{N_r}) \\ \vdots & \vdots & \vdots & \vdots & \vdots \\ x_{N_s}(t_1,\boldsymbol{q}^1) & \cdots & x_{N_s}(t_1,\boldsymbol{q}^r) & \cdots & x_{N_s}(t_1,\boldsymbol{q}^{N_r}) \\ x_1(t_2,\boldsymbol{q}^1) & \cdots & x_1(t_2,\boldsymbol{q}^r) & \cdots & x_1(t_2,\boldsymbol{q}^{N_r}) \\ \vdots & \vdots & \vdots & \vdots & \vdots \\ x_i(t_j,\boldsymbol{q}^1) & \cdots & x_i(t_j,\boldsymbol{q}^r) & \cdots & x_i(t_j,\boldsymbol{q}^{N_r}) \\ \vdots & \vdots & \vdots & \vdots & \vdots \\ x_{N_s}(t_{N_t},\boldsymbol{q}^1) & \cdots & x_{N_s}(t_{N_t},\boldsymbol{q}^r) & \cdots & x_{N_s}(t_{N_t},\boldsymbol{q}^{N_r}) \end{bmatrix}, \tag{3.45}$$

the coefficients matrix

$$\mathbf{B} = \begin{bmatrix} \beta_{1,0}(t_1) & \cdots & \beta_{1,k}(t_1) & \cdots & \beta_{1,N_P}(t_1) \\ \vdots & \vdots & \vdots & \vdots & \vdots \\ \beta_{N_s,0}(t_1) & \cdots & \beta_{N_s,k}(t_1) & \cdots & \beta_{N_s,N_P}(t_1) \\ \beta_{1,0}(t_2) & \cdots & \beta_{1,k}(t_2) & \cdots & \beta_{1,N_P}(t_2) \\ \vdots & \vdots & \vdots & \vdots & \vdots \\ \beta_{i,0}(t_j) & \cdots & \beta_{i,k}(t_j) & \cdots & \beta_{i,N_P}(t_j) \\ \vdots & \vdots & \vdots & \vdots & \vdots \\ \beta_{N_s,0}(t_{N_t}) & \cdots & \beta_{N_s,k}(t_{N_t}) & \cdots & \beta_{N_s,N_P}(t_{N_t}) \end{bmatrix} \tag{3.46}$$

can be computed according to eq. (3.32)

$$\mathbf{B} = \left(\mathbf{Z}^T\mathbf{Z}\right)^{-1}\mathbf{Z}^T\mathbf{X}, \tag{3.47}$$

with the matrix $\mathbf{Z}$ defined by eq. (3.33).

Pseudo discrete projection (NISP)

Regarding to equation (3.36), the coefficients $\beta_k(\boldsymbol{r},t)$ can be calculated by means of the inner products:

$$\beta_k(\boldsymbol{r},t) = \frac{1}{\|\Psi_k(\boldsymbol{Q})\|^2} \int\limits_{\Omega^d} x(\boldsymbol{r},t,\boldsymbol{Q})\Psi_k(\boldsymbol{Q})p_Q(\boldsymbol{q})d\boldsymbol{q} \tag{3.48}$$

$x_k(\boldsymbol{r},t)$ is discretized in the spatial space according to eq. (2.11) and is rewritten in vector form:

$$\boldsymbol{\beta}_k(t) = \frac{1}{\|\Psi_k(\boldsymbol{Q})\|^2} \int\limits_{\Omega^d} \mathbf{x}_k(t,\boldsymbol{Q})\Psi_k(\boldsymbol{Q})p_Q(\boldsymbol{q})d\boldsymbol{q}. \tag{3.49}$$

The multidimensional integration can be computed by means of the sparse grid quadrature (see section 3.2.4):

$$\boldsymbol{\beta}_k(t) = \frac{1}{\|\Psi_k(\boldsymbol{Q})\|^2} \sum_{i=1}^{N_Q} w^i\Psi_k(\boldsymbol{q}^i)\mathbf{x}_k(t,\boldsymbol{q}^i), \tag{3.50}$$

where $\boldsymbol{q}^i$, w^i are the integration nodes and their corresponding weight, while N_Q is the number of integration points. In the implementation, this is achieved by the following matrix multiplication:

$$\mathbf{B} = \underbrace{\mathbf{X}}_{\substack{(N\times N_Q)\\ \text{matrix}}} \cdot \underbrace{\text{diag}\left(\left[w^1,\dots,w^{N_Q}\right]\right)}_{\substack{(N_Q\times N_Q)\\ \text{matrix}}} \cdot \underbrace{\mathbf{Z}}_{\substack{(N_Q\times(N_P+1))\\ \text{matrix}}} \cdot \underbrace{\text{diag}\left(\left[\gamma_0,\dots,\gamma_{N_P}\right]\right)}_{\substack{((N_P+1)\times(N_P+1))\\ \text{matrix}}} \tag{3.51}$$

where

$$\mathbf{Z} = \begin{bmatrix} \Psi_0(\boldsymbol{q}^1) & \cdots & \Psi_k(\boldsymbol{q}^1) & \cdots & \Psi_{N_P}(\boldsymbol{q}^1) \\ \vdots & \ddots & \vdots & \vdots & \vdots \\ \Psi_0(\boldsymbol{q}^i) & \cdots & \Psi_k(\boldsymbol{q}^i) & \cdots & \Psi_{N_P}(\boldsymbol{q}^i) \\ \vdots & \vdots & \vdots & \ddots & \vdots \\ \Psi_0(\boldsymbol{q}^{N_Q}) & \cdots & \Psi_k(\boldsymbol{q}^{N_Q}) & \cdots & \Psi_{N_P}(\boldsymbol{q}^{N_Q}) \end{bmatrix}. \tag{3.52}$$

3.2.3 Discussion of the three computational methods for calculating the gPC coefficients

As there are three computational methods for calculating the gPC coefficients, it is a natural question to ask, which method should be used in practice?

The stochastic Galerkin ensures that the residue of the projection is orthogonal to the space spanned by the gPC basis functions. Its accuracy is also optimal in $\mathcal{L}_2$ sense. The approximation errors of the method are also well investigated. The convergence theory for the stochastic Galerkin method can be found in [Bab04, LM10]. However, the main disadvantage of the stochastic Galerkin method is that the existing deterministic code cannot be used or has to be modified to compute the gPC coefficients.

The non-intrusive methods, namely the stochastic collocation and NISP, on the contrary, are easier to implement. The methods are simple as they need to generate a set of input nodes, running the existing deterministic code of the original problem at the chosen input nodes, and constructing the gPC coefficients.

The stochastic collocation method is a sampling based method, whereas the NISP is a deterministic method. In case of low or moderate number of stochastic dimension d, the NISP is preferable, as the quadrature formula can compute the integration exactly. However, the number of quadrature nodes increases substantially with the stochastic dimension d. Therefore, the stochastic collocation is preferable, in case of a large stochastic dimension d, as the convergence rate of sampling based methods is independent of the stochastic dimension d.

However, there is still a lack of theoretical studies about the error estimation and convergence rate of an approximation in general models for the non-intrusive methods. By the stochastic collocation, an overfitting can occur if the sample size N_r is too large [Cre09]. Similarly, the projection error of NISP method is smaller for higher order of polynomial expansion in many cases, but does not vanish in general. For some problems, the projection error may become critical, if the expansion degree is too high. In fact, for NISP there exists an optimal expansion degree P contributing the lowest projection error [Cre09]. This optimal expansion degree P is unknown in general. The selection of this degree or adaptive strategies to find an optimal gPCE are mostly empirical or based on heuristic rules.

As one of the requirements of the developed framework for analyzing and identification is to avoid new implementations of solver code and the concerned applications deal with small or moderate number of parameters, the NISP is the method mainly applied in this thesis as it is less complex to implement. In practice, an acceptable error control can be achieved without difficulty by the NISP with moderate expansion order in many engineering applications. The main calculation of the NISP is the integration in eq.(3.37), which is accomplished numerically in general. Numerical integration methods are discussed in the following section.

3.2.4 Numerical integration methods

Both Galerkin and pseudospectral projection methods require the integrals over the domain $\Omega \subset \mathbb{R}^d$ in order to compute the coefficients $\boldsymbol{\beta}$ as shown in the equations (3.29) and (3.37) respectively.

The analytical integration for most complex functions is difficult, especially for large stochastic dimension d. Therefore, numerical approximation of integrals is required. Numerical multi-dimensional integration can be found in many fields and various methods have been proposed. Each integration method possesses specific advantages and disadvantages, which have to be taken into account when selecting one of them. In this section, some of the common integration methods usually used in the UQ framework are discussed, namely Monte-Carlo integration, Quasi-Monte-Carlo integration, Tensorization of one-dimensional quadrature formula and Cubature formula based on Smolyak's formula.

Monte-Carlo Integration

The Monte-Carlo method belongs to the stochastic integration techniques. In this method, the integral is approximated from samples $\boldsymbol{\xi}^r = \left[\xi_1^r \ldots \xi_d^r\right]^T$, which are generated from the joint density $p_{\boldsymbol{\xi}}(\boldsymbol{\xi})$. For mutually independent components ξ_α, one can independently sample from the marginal densities $p_{\xi_\alpha}(\xi_\alpha)$. For N_{MC} samples, the Monte-Carlo algorithm approximates the integral with

$$\int_a^b f(\boldsymbol{\xi}) d\boldsymbol{\xi} \approx \frac{1}{N_{\mathrm{MC}}} \sum_{r=1}^{N_{\mathrm{MC}}} f(\boldsymbol{\xi}^r). \tag{3.53}$$

The law of large number ensures that the Monte-Carlo approximation converges to the true value of the integral:

$$\lim_{N_{\mathrm{MC}} \to \infty} \frac{1}{N_{\mathrm{MC}}} \sum_{r=1}^{N_{\mathrm{MC}}} f(\boldsymbol{\xi}^r) = \int_b^a f(\boldsymbol{\xi}) d\boldsymbol{\xi} \tag{3.54}$$

In case of finite N_{MC}, the error of the approximated integral scales with $\frac{1}{\sqrt{N_{\mathrm{MC}}}}$, which does not depend on the number of dimensions d of the integral and the smoothness of the integrand. This is the advantage of Monte-Carlo integration against most deterministic methods that depend on them. The convergence rate $\mathcal{O}\left(\frac{1}{\sqrt{N_{\mathrm{MC}}}}\right)$ is independent of the dimension d, however it is still very low, which is the considerable disadvantage of the method. Using more efficient sampling technique such as Latin Hypercube Sampling [McK00] or Quasi-Monte-Carlo Sampling [Mor95] can improve the convergence rates. However, these techniques are still not competitive in case of low to moderate dimensions, where the deterministic methods, such as tensored or sparse grid techniques, are advantageous.

Deterministic quadrature method: univariate quadrature formulas

The most common technique for deterministic integration is called *quadrature.* The quadrature rule is an approximation of definite integral of a function usually by a weighted sum of function values at specified points, also called *quadrature nodes* in literature, within in the integration domain

$$\int_{-1}^{1} f(\xi)d\xi \approx \mathcal{Q}f = \sum_{i=1}^{N_Q} w^i f(\xi^i), \tag{3.55}$$

where ξ^i and w^i are the integration points and their corresponding weights respectively, while N_Q is the number of integration points. There are several methods how to built a quadrature formula. The well known quadrature formulas are e.g. *Newton-Cotes, Clenshaw-Curtis* or *Gaussian Quadrature.*

The *Newton-Cotes* formulas, also known as *trapezoidal rule,* use equidistant abscissas, which can be useful, if the integrand at equally spaced points is given. The properties for a large number of points N_Q are very bad. The formulas do not converge for a general continuous integrand f.

Comparing to the Newton-Cotes formula the *Clenshaw-Curtis* formulas are numerically more stable [Tre08]. The Clenshaw-Curtis formulas use the roots or the extreme points of the Chebyshev polynomials as integration points. The three main variations of integration points for the Clenshaw-Curtis quadrature are:

- Fejer's "first rule" [Fej33] : Chebyshev roots in $(-1,1)$
- Fejer's "second rule" [Fej33]: Chebyshev extrema in $(-1,1)$
- Clenshaw and Curtis [Cle60]: Chebyshev extrema in $[-1,1]$

The first variant are also called "classical" Clenshaw-Curtis formula whereas the third variant also known as "practical" Clenshaw-Curtis formula [Tre08]. The other variations of Clenshaw-Curtis quadrature can be found in [Not97].

One of the most important properties of quadrature formulas is the *degree of polynomial exactness.* A quadrature formula has polynomial exactness *PE,* if the quadrature $\mathcal{Q}f$ provides the exact value of the integral of f for all f from the polynomials of degree *PE*. Both Newton-Cotes and the Clenshaw-Curtis formulas have the degree of polynomial exactness of $N_Q - 1$.

The *Gaussian Quadrature*, named after Carl Friedrich Gauss, is one of the most significant quadrature formulas. The rule is constructed by approximating the integrand by an interpolating polynomial using a specified number of points. The N_Q-points Gauss quadrature rule has a degree of polynomial exactness equal to $2N_Q - 1$. Depending of the smoothness of the function, the number of the points N_Q can be determined to achieve the exactness or to minimize the integration error. More details and comparison of these quadrature rules can be found in [Tre08].

The integration can be expressed in a more general form by introducing a positive weight function $p(\xi)$ into the integrand and generalizing the interval to $[a,b]$. Such an integral can be computed with an optimal accuracy by the Gaussian quadrature rules, which are constructed by using orthogonal polynomials with the weight $p(\xi)$.

$$\int_b^a f(\xi)p(\xi)d\xi \approx \sum_{i=1}^{N_Q} w^i f(\xi^i) \tag{3.56}$$

This formulation corresponds to the inner product in eq. (3.29) and (3.37), in which the weight function $p(\xi)$ is interpreted as the probability density function $p_\xi(\xi)$. The correspondence between the Gauss quadrature formulas and certain types of pdfs is shown in table 3.3.

Distribution $p(\xi)$	*Interval* $\mathfrak{X}$	*Weight*	*Quadrature*
Gaussian	$(-\infty,\infty)$	$e^{-\xi^2}$	Gauss-Hermite
Gamma	$[0,\infty)$	$e^{-\xi}$ $\xi^\alpha e^{-\xi}$ for $\alpha > -1$	Gauss-Laguerre Generalized Gauss-Laguerre
Beta	$[-1,1]$	$(1-\xi)^\alpha (1+\xi)^\beta$	Gauss-Jacobi
Uniform	$[-1,1]$	1	Gauss-Lengendre, Clenshaw-Curtis or Gauss- Patterson

Table 3.3: Quadrature formula corresponding to certain types of probability distributions

Changes of the interval and the parameters of pdf

The weights and the nodes of the quadrature formula are defined for the distribution with the particularly parameter on the specific support interval, for example $[-1,1]$ in case of Gauss-Legendre quadrature. Computing the integral over arbitrary interval of interest $[a,b]$ needs a transformation:

$$\xi = \chi(\zeta), \tag{3.57}$$

where $\chi(\cdot)$ is the transformation function mapping from the ζ RV defined on the specific support interval $\mathfrak{X}$ to ξ RV defined on the arbitrary interval Ω. The integral yields:

$$\int_{\Omega} f(\xi)p(\xi)d\xi = \int_{\mathfrak{X}} f(\chi(\zeta))p(\chi(\zeta))\left|\frac{d\xi}{d\zeta}\right| d\zeta. \tag{3.58}$$

The elaborated formulas of the transformations and the quadrature for the distributions using in the gPCE can be found in appendix B.

Nested quadrature

For further discussion, the following notation is introduced to distinguish formulas with different accuracy level and extending to multi-dimensional integration:

$$\mathcal{Q}_l^d f = \sum_{i=1}^{N_{\mathcal{Q}_l^d}} w^i f(\boldsymbol{\xi}^i)$$

is the d-dimensional quadrature approximation of integrand f with the accuracy level $l \in \mathbb{N}$. The underlying grid of the quadrature formulas is defined by the notation $\boldsymbol{\Theta}_l^d = \left\{\xi_l^1, \ldots, \xi_l^{N_{\mathcal{Q}_l^d}}\right\}$.

For many applications, e.g. multi-dimensional integration or adaptive quadrature, it is expedient that a quadrature rule is *nested*. A nested quadrature rule is a quadrature rule, for which the node set of each formula is a subset of the node set of its successors.

$$\boldsymbol{\Theta}_l^d \subset \boldsymbol{\Theta}_{l+1}^d$$

For example, the Clenshaw-Curtis quadrature is nested, but the Gauss formulas are in general not nested. Patterson [Pat68] proposed a sequence of nested Gauss quadrature formulas with maximal degree of exactness, by iterating Kronrod's scheme [Kro65] recursively. This quadrature formula is known as *Gauss-Patterson* rule. The differences between the nested quadrature and the normal quadrature are illustrated in figure 3.1. By the *Gauss-Patterson* rule, which is a nested quadrature rule, the nodes of the lower level are always found in the higher level. As the *Gauss-Legendre* is not a nested quadrature, the level in the sense of a sparse grid is not defined. The *Gauss-Legendre* with the same number of quadrature nodes is presented in figure 3.1b. It can be seen that the nodes in higher level is not a subset of the nodes from the lower level, except the middle point, and therefore the *Gauss-Legendre* is not nested.

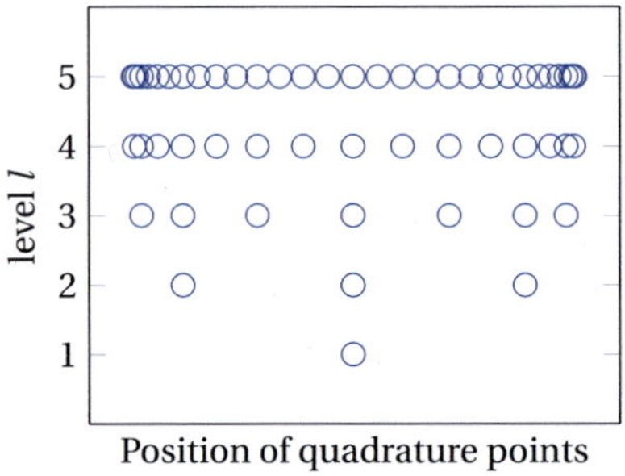

(a) Gauss-Patterson formulas

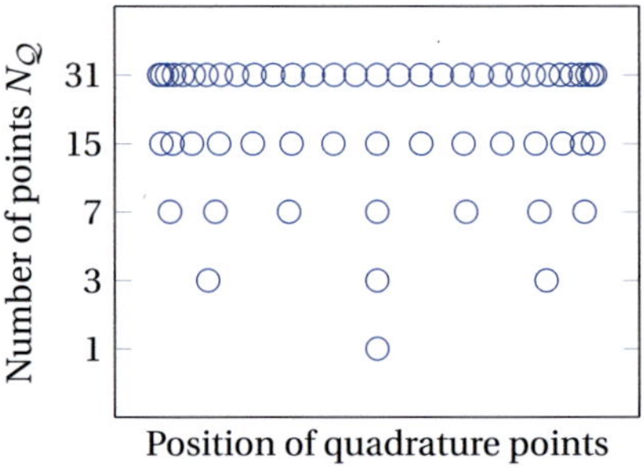

(b) Gauss-Legendre formulas

Figure 3.1: Comparison of integration points of the Gauss-Legendre quadrature (not nested) and Gauss-Patterson quadrature (nested) rule

Multi-dimensional numerical integration: Tensor product formulation

The d-dimensional numerical integration can be constructed by tensorization of the 1D Quadrature formulas (3.56).

$$\begin{aligned} \mathcal{Q}_l^d f &= \left(\mathcal{Q}_{l_1}^1 \otimes \cdots \otimes \mathcal{Q}_{l_d}^1\right) f \\ &= \sum_{i_1=1}^{N_{\mathcal{Q}_l 1}} \cdots \sum_{i_d=1}^{N_{\mathcal{Q}_l d}} w_{l_1}^{i_1} \cdots w_{l_d}^{i_d} \cdot f\left(\xi_1{}^{i_1}, \cdots, \xi_d{}^{i_d}\right). \end{aligned}$$

With this formula, the total number of quadrature points is

$$N_{Q_l^d} = \prod_{\alpha=1}^{d} N_{Q_{l\alpha}}, \tag{3.59}$$

where $N_{Q_{l\alpha}}$ is the number of quadrature point of the α-th dimension. In order to distinguish the formula for the multi-dimensional numerical integration from quadrature formula for the univariate numerical integration, the formula of the multi-dimensional numerical integration is called *cubature* in many literature. As an example, figure 3.2 illustrates the 2D cubature rules constructed by tensored products of Gauss-Patterson quadrature.

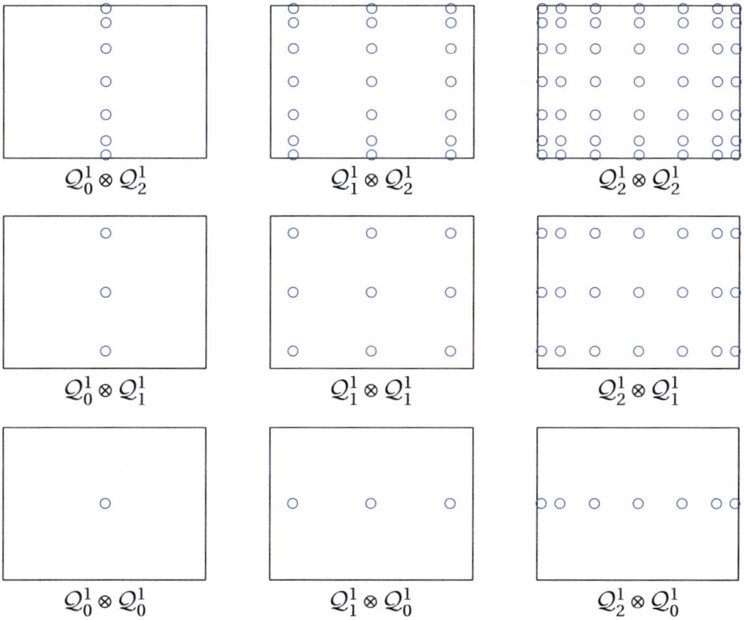

Figure 3.2: Illustration of 2D-cubature rules constructed by tensored products of Gauss-Patterson quadrature

Equation (3.59) exhibits that $N_{Q_l^d}$ increases exponentially with increasing dimension d. The convergence rate of error bounds decreases also significantly as the dimension grows. The exponential growth in the required number of quadrature points and the diminished convergence rate, referred to as *"the curse of dimensionality"*, results that the tensored cubature formula, even if formulated based on the optimal quadrature formulas (Gauss type), is of practical usefulness only for low dimensionality.

In fact, the cubatures resulting from full tensorization are non-optimal as demonstrated in the simple example of an integral in two dimensions shown in figure 3.3. The figure presents the monomials of two-dimensional tensored products of the polynomial of order 5. If the required degree of polynomial exactness (PE) is 5, only the monomials below the blue line are required to approximate the integration. From this simple example, the cubature rules by full tensorization involve higher monomials than necessary (the monomials above the line $PE = 5$). It can be seen that their degree of exactness could be actually achieved using a lower number of nodes. This result motivates the usage of sparse grid quadrature techniques for problems with moderate dimensionality.

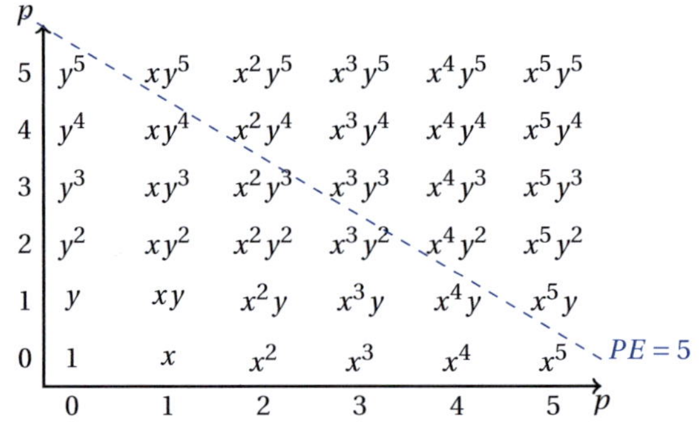

Figure 3.3: Monomials of two-dimensional tensored products of the polynomial of order 5

Sparse grid cubatures

The first *sparse grid* method was proposed by Smolyak [Smo63] in the context of multi-dimensional quadrature and interpolation. Original goal of the sparse grid method is to construct grids and weights that yield the same accuracy as tensor product formulas but with a significantly reduced number of required nodes.

The numerical integrations with sparse grids are proposed by several authors. Novak and Ritter proposed the application of sparse grids to numerical integration using the the Clenshaw-Curtis rule [Nov96]. Gerstner and Gribel concluded that the sparse grid with Gauss-Patterson formulas perform best in comparison to Gauss-Legendre, Clenshaw-Curtis or trapezoidal formulas by considering the ratio of necessary function evaluations and accuracy [Ger98].

Sparse grid construction

Consider a family of 1D-quadrature rules $\mathcal{Q}_l^1 f$ as in eq. (3.55) with the 1D nodal points set $\boldsymbol{\Theta}_l^1 = \{\xi_l^1, \dots, \xi_l^n\}$. The sparse grid $\mathcal{S}(L,d)$ for level L and dimension d is:

$$\mathcal{S}(L,d) = \sum_{L-d+1 \le |\underline{l}| \le L} (-1)^{L-|\underline{l}|} \cdot \begin{pmatrix} d-1 \\ L-|\underline{l}| \end{pmatrix} \left(\mathcal{Q}_{l_1}^1 \otimes \cdots \otimes \mathcal{Q}_{l_d}^1 \right), \tag{3.60}$$

where $\underline{l}$ is the multi-index level of the component rules used and $|\underline{l}| = \sum_{j=1}^{d} l_j$ and the nodal set for the sparse grid is

$$\mathcal{H}(L,d) = \bigcup_{L-d+1 \le |\underline{l}| \le L} \boldsymbol{\Theta}_{l_1}^1 \times \cdots \times \boldsymbol{\Theta}_{l_d}^1. \tag{3.61}$$

To demonstrate the formula, an example is presented here. Using eq. (3.60) the two-dimensional, level 2 sparse grid $\mathcal{S}(2,2)$ is constructed as

$$\mathcal{S}(2,2) = \left(\mathcal{Q}_2^1 \otimes \mathcal{Q}_0^1\right) + \left(\mathcal{Q}_0^1 \otimes \mathcal{Q}_2^1\right) + \left(\mathcal{Q}_1^1 \otimes \mathcal{Q}_1^1\right) - \left(\mathcal{Q}_0^1 \otimes \mathcal{Q}_1^1\right) - \left(\mathcal{Q}_1^1 \otimes \mathcal{Q}_0^1\right).$$

The nodes of the sparse grid $\mathcal{S}(2,2)$ are presented in figure 3.4. The sparse grid $\mathcal{S}(2,2)$ has 17 nodes $(7+7+9-3-3)$ compared to 49 nodes (7×7) of the tensored product quadrature formula $\mathcal{Q}_2^1 \otimes \mathcal{Q}_2^1$ (cf. figure 3.2).

In this thesis, the calculation of nodes and weights of sparse grid quadrature is based on the implementation of *Sparse Grid Mixed Growth Anisotropic Rules* by Burkardt [Bur]. The reader is referred to [Bur, Nov96, Ger98] for more information about sparse grids and calculation of their weights. The discussion about the usage of sparse grids in UQ can be found in [LM10, Smi13, Xiu05].

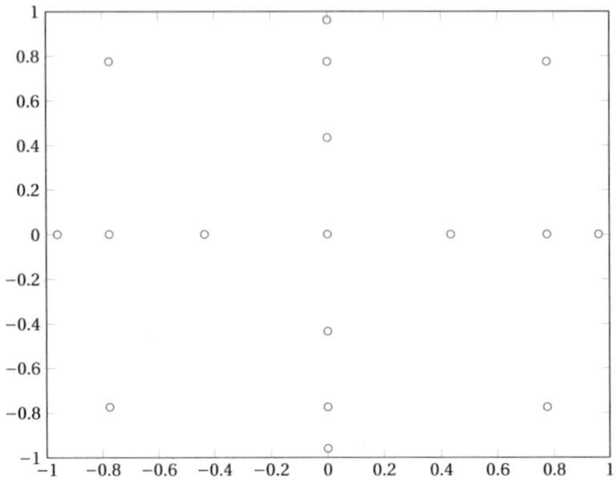

Figure 3.4: Nodes of the two-dimensional, level 2 sparse grid $\mathcal{S}(2,2)$

3.3 Software for the gPC framework

Analyzing the system with UQ allows one to see new aspects of the system behavior. The pdf and the statistical quantities of the system responses are very useful information for the user to understand the system behavior and optimize plants and the control strategies. In practice, the model is often implemented in the form of a deterministic solver program. As mentioned in the chapter 1, the sampling-based methods such as Monte-Carlo are generally adopted to assess the pdf of the system responses. The methods are straightforward by solving the model $\mathcal{M}_s\left(x(\boldsymbol{r},t),\boldsymbol{q}\right)$ for input realizations from the distribution $p(\boldsymbol{q})$, which normally needs extensive computation.

The gPC is an efficient approach to approximate the pdf and the statistical moments of the system responses. As discussed in section 3.1, the statistical moments can be determined directly from the coefficients. The pdf of the system responses can be approximated by using the gPC as surrogate model, which is computationally cheaper than a full model computation.

The implementation of gPC consists of selecting appropriate orthogonal polynomials and computing their corresponding coefficients. Regarding the pdf of the uncertain parameters (see table 3.4), the orthogonal polynomials are established by the equation (3.8). The corresponding coefficients are calculated as discussed in section 3.2.1.

In the course of the thesis, a gPC software package for the framework has been developed. In order to limit the developing effort and avoid the modification of existing deterministic solver code, one of the primary requirement of the software package is that it can work with any already existing deterministic solver. Hence, the non-intrusive approach is adopted in the implementation as discussed in section 3.2.3.

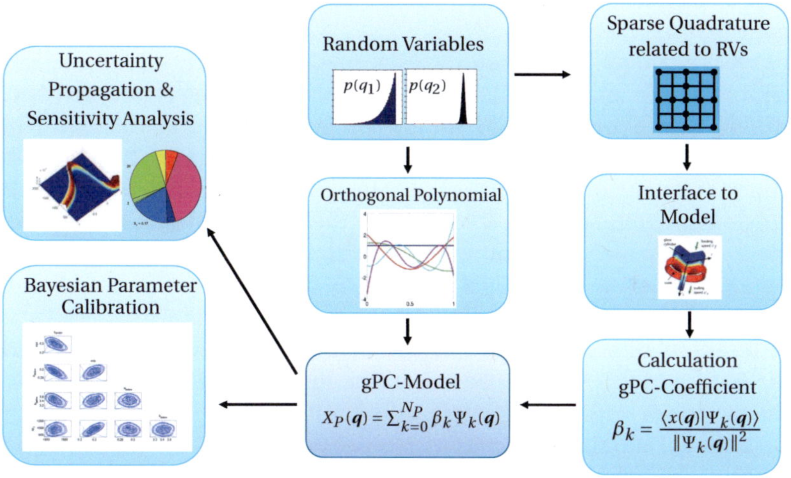

Figure 3.5: Scheme of the software system for gPC framework

The software package is implemented under the MATLAB environment in form of a toolbox with visualization functions. The interaction between the gPC-toolbox and third party programs (i.e. existing FEM, CFD code etc.) is done by the manipulation of the input and output interface of the deterministic solver code.

The scheme of the software system is illustrated in figure 3.5. The procedure begins with defining the parameter uncertainties. In the current software version, the parameters are assumed to be independent. The parameter uncertainty is defined by the marginal probability density function of each parameter. The current software version supports four types of pdfs regarding the gPCE, namely uniform, normal, gamma and beta distribution. The software user sets the type of pdf of each parameter with the corresponding pdf parameters in the software. The marginal distributions are labeled with the parameter name, so that the software interfaces with third party programs.

The multi-dimensional gPC orthogonal basis polynomials are constructed as stated in (3.8). The one-dimensional polynomials are determined based on the pdf of each uncertain parameter regarding to the table 3.4. The polynomial order P is decided by the user. The other types of basis functions such as wavelets or Padé-Legendre are also possible, but they are out of scope of the thesis.

The software allows two non-intrusive methods, namely the least-squares approximation and the NISP method. According to the concerned applications, where the stochastic dimension is small or moderate, the NISP method is preferable because of its deterministic nature. The level of accuracy of the quadrature formula can be determined directly from the expansion polynomial order. In order that the quadrature can compute the integral (3.37) exactly, the program determines the level of the quadrature formula, so that the polynomial exactness of the quadrature formula is more than $2P$.

The calculation of integration points and weights for the sparse grid quadrature in this thesis is based on the implementation of *Sparse Grid Mixed Growth Anisotropic Rules* by Burkardt [Bur]. The system responses of the model and the input parameters are predefined in the interface part between the software and the third party programs. The software runs the deterministic solver at the transformed quadrature nodes. Then, the gPC coefficients are computed by NISP method described in section 3.2.2. All correspondences between the information about the parameters, the probability distributions according to the *Maximum Entropy Principle*, the gPC basis polynomials and the quadrature rules used in the developed framework are summarized in table 3.4.

Information about q_i	*MaxEnt pdf $p_{Q_i}(q_i)$*	*gPC basis Polynomial*	*Quadrature*
mean μ, standard deviation σ	Gaussian	Hermite	Gauss-Hermite
minimum a, maximum b	Uniform	Legendre	Gauss-Patterson
$q_i > 0$, mean μ	Gamma	Laguerre	Gauss-Laguerre

Table 3.4: Correspondence between the information about the parameters, the probability distributions according to the Maximum Entropy Principle, the gPC basis polynomials and the quadrature rules

As the gPC approximation, consisting of the orthogonal polynomials and the corresponding coefficients, is obtained, the system analysis can be conducted. The gPC approximation can be used to assess the pdf of the quantities of interest. The sensitivity indices can be computed from the gPC coefficients to find the influence of each parameter. The sensitivity analysis with gPC is discussed in chapter 4. The gPC can be also used as a surrogate model, e.g. for the Bayesian inverse problem solving as discussed in chapter 5. In the following section, the application of gPC to the uncertainty propagation is presented.

3.4 Performance evaluation of uncertainty propagation using gPC by means of a numerical example

To illustrate the computation method using gPC, the gPCE is applied to the neutron diffusion equation from [Cac14a] as an example. The equation describes neutron diffusion in a one-dimensional pool of water containing distributed neutron sources, as would be typical for a long and deep spent-fuel storage pool. Cacuci conducted the sensitivity analysis to the model and uses the sensitivity to calibrate the parameters. In this chapter, the gPCE is performed to compute the uncertainty propagation. The sensitivity analysis and the parameter calibration of the example are discussed in chapters 4 and 5 respectively.

Considering the diffusion of monoenergetic neutrons due to uniform distributed sources of strength S neutrons/cm$^3 \cdot$ s within a slab of material with thickness $2a$, this reactor problem can be described mathematically with the differential equation

$$\Sigma_a \varphi - D \frac{\partial^2 \varphi}{\partial x^2} = S, \quad x \in (-a, a), \tag{3.62}$$

where $\varphi(x)$ denotes the neutron flux, D is the diffusion coefficient, Σ_a denotes the macroscopic absorption cross section, and S is the distributed source term. The boundary conditions are:

$$\varphi(\pm a) = 0. \tag{3.63}$$

The typical system response for the neutron diffusion problem is the number of neutrons measured by the detector of width Σ_d located at $x = b$

$$y = \Sigma_d \varphi(b). \tag{3.64}$$

Equation (3.62) is considered as a special case of the convection-diffusion equation (2.6) by considering the steady state with zero velocity $\boldsymbol{v} = 0$.

The relevant parameters in this example are collected in the parameter vector

$$\boldsymbol{q} = [\Sigma_a, D, S, \Sigma_d]^T. \tag{3.65}$$

The analytical solution of the BVP is:

$$\varphi(x) = \frac{S}{\Sigma_a}\left(1 - \frac{\cosh(xk)}{\cosh(ak)}\right), \quad k = \sqrt{\Sigma_a / D}. \tag{3.66}$$

It is noted that the analytical solution for DPSs is not available in general. The example with the available analytical solution is chosen, in order that it is possible to evaluate the method. The parameters in the example are considered to be uncertain with the following data: The nominal values of the parameter are $\Sigma_a^0 = 0.0197\ \mathrm{cm}^{-1}$, $D^0 = 0.16$ cm, $S^0 = 10^7$neutrons $\mathrm{cm}^{-3}\mathrm{s}^{-1}$ and $\Sigma_d^0 = 7.438\ \mathrm{cm}^{-1}$. The uncertainties are given by the following relative standard deviations: $\Delta\Sigma_a^0/\Sigma_a^0 = 5\%$, $\Delta S^0/S^0 = 15\%$, $\Delta D^0/D^0 = 5\%$ and $\Delta\Sigma_d^0/\Sigma_d^0 = 10\%$. All parameters are assumed to be independent.

In the developed framework, the uncertain parameters are considered as the random vector. Regarding the maximum entropy principle (see section 2.1.5), the pdf of the RV is normal distribution as follows:

$$\boldsymbol{Q} \sim \mathcal{N}\left(\boldsymbol{q}^0, \mathbf{C}_{\boldsymbol{Q}}\right)$$

$$\boldsymbol{q}^0 = \begin{bmatrix}0.0197 & 0.16 & 10^7 & 7.438\end{bmatrix}^T$$

$$\mathbf{C}_{\boldsymbol{Q}} = \begin{bmatrix} (9.85\times10^{-5})^2 & 0 & 0 & 0 \\ 0 & (8.0\times10^{-3})^2 & 0 & 0 \\ 0 & 0 & (1.5\times10^{6})^2 & 0 \\ 0 & 0 & 0 & (7.44\times10^{-1})^2 \end{bmatrix}$$

The sampling-based method of the uncertainty propagation is achieved by sampling N_{samp} realizations from $\mathcal{N}\left(\boldsymbol{q}^0, \mathbf{C}_{\boldsymbol{Q}}\right)$ and then solving the equations (3.62) and (3.64) for all realizations to assess the pdf of the system responses.

The mean values and the 2σ intervals of the sampling-based uncertainty propagation results with different numbers of realizations are presented in figure 3.6. The figure shows that the results depend on the sample number and a large number of realizations is required in order to achieve the good approximated solution. Such a UQ computation with a sampling-based method is only affordable in case of a model with a relatively low computational complexity.

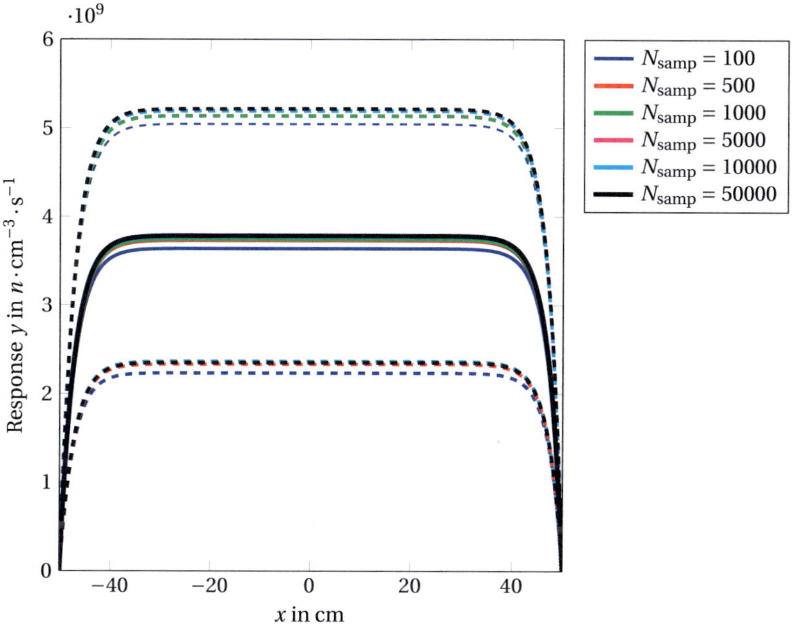

Figure 3.6: Uncertainty of the neutron response along the x-coordinate, the mean values (solid lines) and the 2σ intervals (dashed lines) calculated by the sampling-based method

In contrast, the number of model calculations with the gPC method depends on the number of the quadrature points. The number of the quadrature points can be determined from the desired polynomial exactness of the integral of equation (3.37). In order to calculate the integral $\int y(\boldsymbol{\xi})\Psi_k(\boldsymbol{\xi})p_\xi(\boldsymbol{\xi})d\boldsymbol{\xi}$, the polynomial exactness of the quadrature rule must be more than the polynomial order of the integrand $y(\boldsymbol{\xi})\Psi_k(\boldsymbol{\xi})$. When the system response is approximated by the gPC of order P, the appropriate approximation of the integrand is a polynomial of degree $2P$. Table 3.5 demonstrates the node number and the polynomial exactness of the univariate Gauss-Hermite quadrature (GHQ) rule.

Accuracy level l	0	1	2	3	4	5
Number of univariate GHQ nodes $N_l^1 = 2l+1$	1	3	5	7	9	11
Polynomial exactness of GHQ rule $2N_l^1 - 1$	1	5	9	13	17	21

Table 3.5: Number of nodes and the polynomial exactness of the univariate Gauss-Hermite Quadrature (GHQ) rule with respect to the level of accuracy

For multi-dimensional integrals, the sparse grid quadrature is more efficient, but the sparse grid construction requires a nested quadrature rule, whereas GHQs are not nested. However, GHQ is known to be accurate so that there can be tradeoff. Burkardt's algorithm [Bur] constructs a GHQ sparse grid by grouping nearby nodes as a weighted node.

In this example, the gPC approximations of various polynomial orders are computed. The number of the gPC coefficients regarding to the polynomial order and the required sparse grid level including their node numbers are presented in table 3.6.

Order of Polynomial expansion$P$	0	1	2	3	4	5	6	7	8
Number of gPC coefficients	1	5	15	35	70	126	210	330	495
Required polynomial exactness ($2P$)	0	2	4	6	8	10	12	14	16
Sparse grid level	0	1	1	2	2	3	3	4	4
Polynomial exactness of sparse grid	1	5	5	9	9	13	13	17	17
Number of sparse grid GHQ nodes	1	9	9	49	49	201	201	681	681

Table 3.6: Number of the gPC coefficients and the required polynomial exactness of quadrature rules in order to compute the coefficients with respect to the polynomial expansion order, compared with the number of sparse grid nodes of Gauss-Hermite Quadrature (GHQ) and their polynomial exactness with respect to the sparse grid level

The mean values and 2σ interval resulted by different gPC order are demonstrated in figure 3.7. The mean value and the standard deviation can be computed from the gPC coefficients using equations (3.14) and (3.15). The difference between the different polynomial order approximations are so small that they cannot be observed here. The results of the gPC method are also in extremely good agreement with the result of the sampling-based method with a large sample number. This shows how efficient the gPC approximation is. The gPC method with sparse grid requires less than hundred model calculations and can calculate an equivalent result as from the sampling-based method with more than 10000 model calculations in order to achieve the required accuracy.

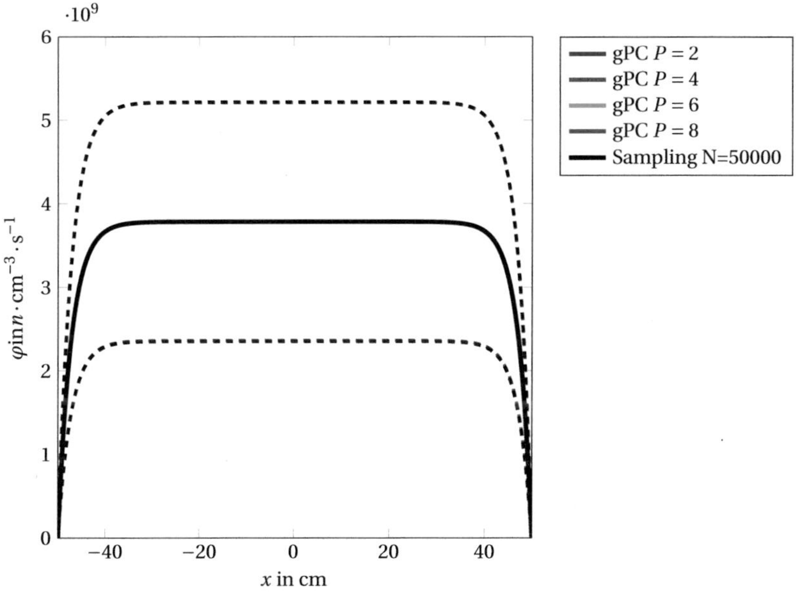

Figure 3.7: Uncertainty of the response along the x-coordinate, the mean values (solid lines) and the 2σ intervals (dashed lines) calculated by gPC

The mean values and the standard deviations of the system response $y = \Sigma_d \varphi(10\text{cm})$ for all tests are shown in table 3.7. The result by MC-simulation with $N_{\text{samp}} = 10^6$ serves as reference. It can be seen that the gPC can approximate the mean value accurately with only the polynomial order of $P = 2$, whereas the sampling-based method needs at least sample size of $N_{\text{samp}} = 50000$ in order to achieve the same level of accuracy. For the standard deviation, the gPC with $P = 8$, which needs to solve the model only 681 times, provides the same accuracy level of the sampling-based method with $N_{\text{samp}} = 10^6$. It should be noted that the results of the sampling-based method for a low sample numbers vary depending on the sampled realizations. The method in [Cac14a] is a local expansion-based method, which expands the system response to first-order Taylor series by using the Jacobi matrix. As it is a linear approximation, the method result in a Gauss distribution determined by the mean and the standard deviation. The mean value and the standard deviation of Cacuci's method has the accuracy level about the accuracy of the sampling-based method with $N_{\text{samp}} = 10000$.

Methods	mean	standard deviation
MC $N_{\text{samp}} = 100$	3.694×10^9	7.030×10^8
MC $N_{\text{samp}} = 500$	3.735×10^9	7.011×10^8
MC $N_{\text{samp}} = 1000$	3.750×10^9	6.925×10^8
MC $N_{\text{samp}} = 5000$	3.778×10^9	7.088×10^8
MC $N_{\text{samp}} = 10000$	3.777×10^9	7.075×10^8
MC $N_{\text{samp}} = 50000$	3.784×10^9	7.161×10^8
MC $N_{\text{samp}} = 1000000$	3.785×10^9	7.115×10^8
gPC $P = 2$	3.785×10^9	7.070×10^8
gPC $P = 4$	3.785×10^9	7.147×10^8
gPC $P = 6$	3.785×10^9	7.124×10^8
gPC $P = 8$	3.785×10^9	7.116×10^8
Method in [Cac14a]	3.77×10^9	7.057×10^8

Table 3.7: The mean and the standard deviation results by the Monte-Carlo (MC) method with different sampling numbers N_{samp} compared to the generalized polynomial chaos (gPC) method with different expansion orders P and the result in [Cac14a]

Figure 3.8 shows the approximation of pdf of response $y = \Sigma_d \varphi(x = 10\text{cm})$ by the gPC method, the sampling-based method and the method in [Cac14a]. The histogram obtained by MC-simulation with $N_{\text{samp}} = 10^6$ should serve as reference pdf. The Gauss distribution regarding the mean and standard deviation from [Cac14a] is also illustrated for comparison.

The gPC of order $P = 8$ is noticeably accurate regardless of much lower computational effort compared to the sampling based method. As shown in table 3.6, the gPC of order $P = 8$ required only 681 evaluations. The sampling-based method with similar effort, presented by the result of MC with $N_{\text{samp}} = 10^3$, cannot provide the comparable result.

From this example, one could see the potential of the gPC method. The gPC approximation requires a much lower number of evaluations. The method is very useful in case of an extensive computational model, where a large number of evaluations is not acceptable. The local expansion-based method proposed in [Cac14a, Cac14b, Bad12, Ars14] does not need repeated calculations. However, the method requires derivatives of system responses to establish a Jacobi matrix, which could be difficult, when analytical solutions are not available. This issue is related to the sensitivity analysis, which is discussed in the next chapter.

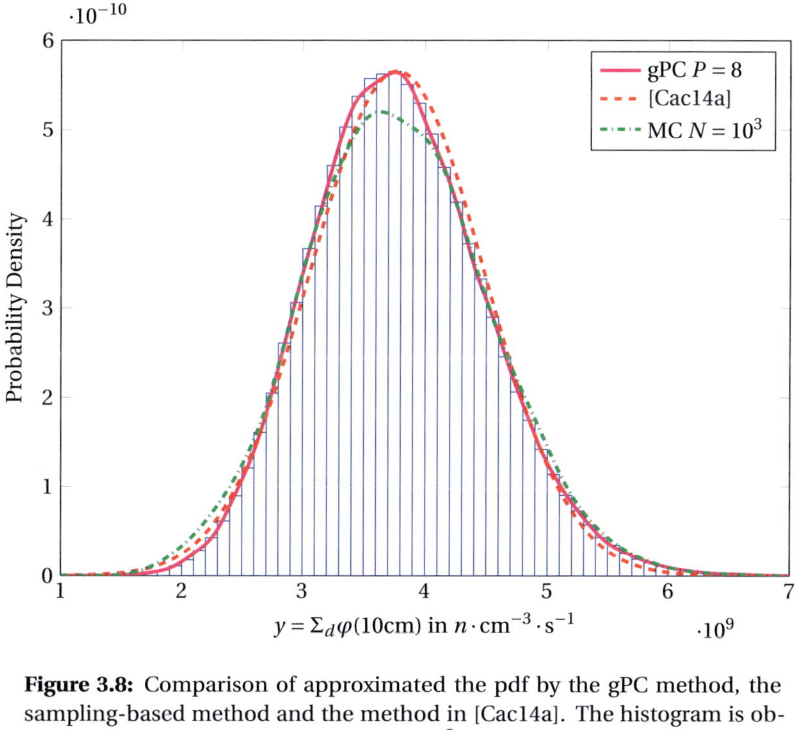

Figure 3.8: Comparison of approximated the pdf by the gPC method, the sampling-based method and the method in [Cac14a]. The histogram is obtained by MC-simulation with $N_{\text{samp}} = 10^6$ and serves as a reference for the pobability density function.

4

Sensitivity Analysis

Comprehension of the intrinsic relationships between model inputs, model outputs and model parameters is an important information for the model user. Knowing the relative uncertainty in a model with different sets of parameters and input data should contribute to a better understanding of the relationships between model assumptions, parameters, inputs, and model predictions. *Sensitivity Analysis* (SA) is the study of how the uncertainty in the system responses can be apportioned to different sources of the input uncertainties.

The term *Sensitivity analysis* has different connotations in various scientific communities. Morgan et al. [Mor92] give a broad definition of sensitivity analysis as the determination of how a change in any aspect of the model changes any predicted model output. According to the definition, the meaning of the *change* of the model could be very broad from the specification of the system and the surrounding to the model parameters. In this thesis the model changes due to the parameter changes is considered, and the model output is mathematically defined as an operator on the system state vector as mentioned in section 2.1.3. Therefore, the objective of sensitivity analysis can be viewed as quantifying the *relative* contributions due to individual parameters and determining how variations in parameters affect the system responses.

Apart from improving the understanding of the system, the information about the sensitivity of the model parameters are most commonly used in system design, optimization, and model calibration. For example, one could consider how all parameters affect the performance or reliability of the system. Some insensitive parameters can be fixed in the optimization of model calibration since their variation minimally influences outputs.

The methods for sensitivity analysis can be classified in two groups as local and global sensitivity analysis. The objective of *local* SA is to analyze the behavior of the system responses locally around a chosen point or trajectory. On the other hand, the objective of *global* SA is to determine all the global variation of a system response over the entire range of input values. Many methods for both local and global SA have been proposed; however, applying SA to the DPS have some challenges w.r.t. computation. This chapter gives a brief introduction to methods of both groups. Then, the challenges of the sensitivity analysis computation for DPS are discussed and their solutions are also proposed with numerical examples.

4.1 Local sensitivity analysis

Cacuci [Cac03] gives a description of *Local Sensitivity Analysis (LSA)* as analyzing the behavior of the system responses locally around a chosen point or trajectory in the combined phase space of parameters and state variables. Mostly, the chosen point or the chosen trajectory are a nominal value or nominal trajectory. The local sensitivity is defined as a local measure of given inputs or parameters on determined outputs. This is typically achieved by the derivative of the response with respect to each of the individual parameters q_i.

The definition of a local sensitivity S_i of an output variable y to a parameter q_i is defined by the derivative of y with respect to the variable q_i at the nominal value q_i^0 by the equation

$$S_i = \left.\frac{\partial y}{\partial q_i}\right|_{q_i=q_i^0}. \tag{4.1}$$

In simple words, the local sensitivity is the slope around a chosen point or a chosen trajectory.

In case of the n-dimensional vector output $\boldsymbol{y}$ and the d-dimensional vector parameter $\boldsymbol{q}$, the derivative of the system response is described by the **Jacobi Matrix**. In the course of sensitivity analysis, this matrix is sometimes called **sensitivity matrix**. The Jacobi matrix is the matrix of all first-order partial derivatives of a vector-valued function.

$$\mathbf{J} = \begin{bmatrix} \frac{\partial y_1}{\partial q_1} & \cdots & \frac{\partial y_1}{\partial q_d} \\ \vdots & \ddots & \vdots \\ \frac{\partial y_n}{\partial q_1} & \cdots & \frac{\partial y_n}{\partial q_d} \end{bmatrix} \tag{4.2}$$

In case of DPS, the perturbation can be spatially distributed, The derivative (4.1) should be generalized by directional derivatives. Using the Gâteaux derivative, the local sensitivity of a system output is defined as:

$$\delta \boldsymbol{y} = \lim_{\varepsilon \to 0} \frac{\boldsymbol{y}(\boldsymbol{q}^0 + \epsilon \boldsymbol{h}) - \boldsymbol{y}(\boldsymbol{q}^0)}{\epsilon}, \tag{4.3}$$

where the variation $\boldsymbol{h}$ defines how the system is varied.

The sensitivity (4.3) can be calculated using three techniques: (a) finite difference approximation, (b) automatic differentiation, or (c) solving the sensitivity equation.

The most elementary is to approximate the derivative using the finite difference relations:

$$\frac{\partial \boldsymbol{y}(\boldsymbol{q})}{\partial \boldsymbol{q}} \approx \begin{cases} \frac{\boldsymbol{y}(\boldsymbol{q}+\epsilon\boldsymbol{h})-\boldsymbol{y}(\boldsymbol{q})}{\epsilon} & \text{forward difference} \\ \frac{\boldsymbol{y}(\boldsymbol{q}+\frac{\epsilon}{2}\boldsymbol{h})-\boldsymbol{y}(\boldsymbol{q}-\frac{\epsilon}{2}\boldsymbol{h})}{\epsilon} & \text{central difference} \\ \frac{\boldsymbol{y}(\boldsymbol{q})-\boldsymbol{y}(\boldsymbol{q}-\epsilon\boldsymbol{h})}{\varepsilon} & \text{backward difference} \end{cases} \tag{4.4}$$

However, this method has the difficulty that the accuracy of (4.4) is highly dependent on the choice of ϵ, which also must be correctly scaled according to the magnitude of $\boldsymbol{q}$. Cacuci suggests using the $\delta \boldsymbol{q} = \epsilon \boldsymbol{h}$ of the order of 1 % from its nominal values $\boldsymbol{q}^0$ [Cac03].

For certain problems, automatic differentiation (AD) can be used to calculate the sensitivity. This approach exploits the fact that every computer program executes a sequence of elementary arithmetic operations (e.g. addition, subtraction, multiplication and division) and elementary functions (exponential, logarithms, etc.). AD decomposes the complex operator into a combination of elementary arithmetic operations and function evaluations and then computes the derivatives of arbitrary order by applying the chain rule to these operations. More detail about the automatic differentiation and its implementation can be found in [Nei10].

Another option is to calculate the sensitivity $\delta \boldsymbol{y}$ from the sensitivity equation. By using the Gâteaux variation with the perturbation method, the sensitivity equation can be formulated from the system equation. Solving the sensitivity equation yields the sensitivity $\delta \boldsymbol{y}$ as result. Cacuci proposed an approach named *Forward Sensitivity Analysis Procedure (FSAP)*. In case that the number of the considered parameters is higher than the number of the model outputs, the extensive computation of the forward model can be relieved by formulating the adjoint operator and use it to calculate the sensitivity. This method is known as *Adjoint Sensitivity Analysis Procedure (ASAP)*. More detail about these two methods are discussed explicitly in [Cac03], [Cac05].

For the AD and sensitivity equation techniques, the closed form of the solution is required to compute the sensitivity. In most practical cases of DPS, the analytical closed form solution is not available as mentioned in the chapter 2. The approximation with the finite difference method has its drawback of the dependence on the choice of ϵ. As one of the contributions of this thesis, the computation of the sensitivity by means of the gPC approximation is proposed, which is discussed in the following.

4.2 Computation the local sensitivity by the gPC approximation

As discussed in the chapter 3, the gPC approximation can be used as a surrogate model. The main idea of this approximation method is using the gPC as the response surface of the output regarding to the parameters. The system responses are assumed smooth with respect to the parameter space, which should not be a crucial limitation w.r.t. the DPSs considered in this work. Using the orthogonality properties of the polynomials, the derivative of the system response can be derived from the derivative of the orthogonal polynomials.

Given the DPS with the system response $\boldsymbol{y}$ depending on the parameter vector $\boldsymbol{q}$, each parameter q_i in the parameter vector $\boldsymbol{q}$ is uncertain and assumed to be distributed uniformly in its uncertain range, given by the model user. Then, the P order gPC approximation of the considered system responses $\boldsymbol{y}$ with respect to the considered d-dimensional parameter vector $\boldsymbol{q}$ is constructed.

$$\boldsymbol{y}_P(\boldsymbol{q}) = \sum_{|\underline{i}|=0}^{N_P} \boldsymbol{\beta}_{\underline{i}} \Psi_{\underline{i}}(q_1, q_2, \ldots, q_d) \tag{4.5}$$

The multi-dimensional PC basis functions $\Psi_{\underline{i}}$ are constructed from the 1D-Legendre polynomials according to equation (3.9). The coefficients vector $\boldsymbol{\beta}_{\underline{i}}$ are calculated with the NISP method as mentioned in section 3.2. The local sensitivity analysis method involves taking the partial derivatives of system response with respect to a parameter q_a at some fixed point $\boldsymbol{q}^0$ in the parameter space. This derivative indicates the sensitivity to the parameter q_a at fixed point $\boldsymbol{q}^{\mathbf{0}}$.

Using gPC approximation, the partial derivative with respect to an input variable q_a can be approximated as follows:

$$\frac{\partial \boldsymbol{y}_P}{\partial q_a}(\boldsymbol{q}^0) = \sum_{\underline{i} \in \mathcal{A}_a^T} \boldsymbol{\beta}_{\underline{i}} \left. \frac{\partial \Psi_{\underline{i}}(\boldsymbol{q})}{\partial q_a} \right|_{\boldsymbol{q}^0}, \tag{4.6}$$

where $\mathcal{A}$ is the set of multi-indices $\underline{i}$ of the truncated expansion, $\mathcal{A}_a^T := \{\underline{i} \in \mathcal{A}, i_a > 0\}$ is the subset of $\mathcal{A}$, where i_a is larger than zero. The derivative of the terms $\boldsymbol{\beta}_{\underline{i}} \Psi_{\underline{i}}(\boldsymbol{q})$ with $i_a = 0$ is equal to zero, because they do not depend on the parameter q_a.

To illustrate the computation of sensitivity by the gPC approximation, this method is applied to the available example of the conventional LSA from the Cacuci's work [Cac14a]. In the paper, Cacuci applies his LSA method to the neutron diffusion equation and uses the sensitivity to calibrate the parameter. The details about the parameter calibration related the example will be discussed in the chapter 5.

4.2.1 Performance evaluation of local sensitivity analysis computation using gPC by means of numerical example

In this subsection, the proposed gPC approach to compute the sensitivity is applied to the example described in 3.4. As the analytical solution is available in this case, the local sensitivities can be calculated directly by the partial derivatives. The analytical expression of sensitivities w.r.t. each parameter at the nominal parameter $\boldsymbol{q}^0 = [\Sigma_a^0, D^0, S^0, \Sigma_d^0]^T$ are [Cac14a]:

$$\frac{\partial y}{\partial S} = \frac{\Sigma_d^0}{\Sigma_a^0}\left(1 - \frac{\cosh(bk)}{\cosh(ak)}\right), \tag{4.7}$$

$$\frac{\partial y}{\partial \Sigma_d} = \frac{S^0}{\Sigma_d^0}\left(1 - \frac{\cosh(bk)}{\cosh(ak)}\right), \tag{4.8}$$

$$\frac{\partial y}{\partial \Sigma_a} = -\frac{S^0\Sigma_d^0}{(\Sigma_a^0)^2}\left(1 - \frac{\cosh(bk)}{\cosh(ak)}\right) + \frac{1}{2\sqrt{D^0\Sigma_d^0}}\frac{S^0\Sigma_d^0}{\Sigma_d^0}C, \tag{4.9}$$

$$\frac{\partial y}{\partial D} = \frac{1}{2}\sqrt{\frac{\Sigma_a^0}{D^0}}\frac{S^0\Sigma_d^0}{D^0\sigma_a^0}C, \tag{4.10}$$

with the constant C:

$$C = \frac{a\sinh(ak)\cosh(bk) - b\sinh(bk)\cosh(ak)}{(\cosh(ak))^2}.$$

The system response remains the neutron flux φ at the position $x = 10$ cm. The gPC approximation of the model is constructed in the same way as the example in section 3.4, with the difference, that the parameter uncertainties are determined to be uniformly distributed in the range $[\boldsymbol{q}_0 - 3\Delta\boldsymbol{q}, \boldsymbol{q}_0 + 3\Delta\boldsymbol{q}]$, where $\Delta\boldsymbol{q} = [\Delta\Sigma_a^0, \Delta D^0, \Delta S^0, \Delta\Sigma_d^0]^T$ (cf. section 3.4). The gPC approximation of the uniform distribution possesses naturally the *Gibbs artifact* at the bounds of the interval. Therefore, the intervals of the parameter space is expanded to $3\Delta\boldsymbol{q}$, although the gPC approximation is intentionally applied in the study range $[\boldsymbol{q}_0 - \Delta\boldsymbol{q}, \boldsymbol{q}_0 + \Delta\boldsymbol{q}]$ in order to avoid the *Gibbs phenomenon* at the bounds. Then, the gPC approximations of the derivatives with respect to each parameter is computed by equation 4.6.

As the gPC approximations of the derivatives with respect to each parameter result in a similar manner, the sensitivity of the parameter Σ_a is discussed here as an instance. Figure 4.1 shows the comparison of the gPC approximation of the sensitivity of the parameter Σ_a in the interval $[\Sigma_a^0 \pm \Delta\Sigma_a^0]$. Figure 4.1a reveals the agreement between the analytic solution and the gPC approximation of order $P = 4$ and $P = 7$ of the sensitivity. To evaluate the approximation for all parameters, the following relative error is used as a measure of the approximation:

$$e_{\text{rel}} := \frac{\left|S_{\text{gPC}} - S_{\text{an}}\right|}{|S_{\text{an}}|}. \tag{4.11}$$

Figure 4.1b shows the relative errors between the approximation and analytical values. The gPC $P = 7$ provide a very accurate approximation with the relative error about 10^{-7}. The gPC $P = 4$ has a lower accuracy, but it is also required a lower number of model evaluation to construct.

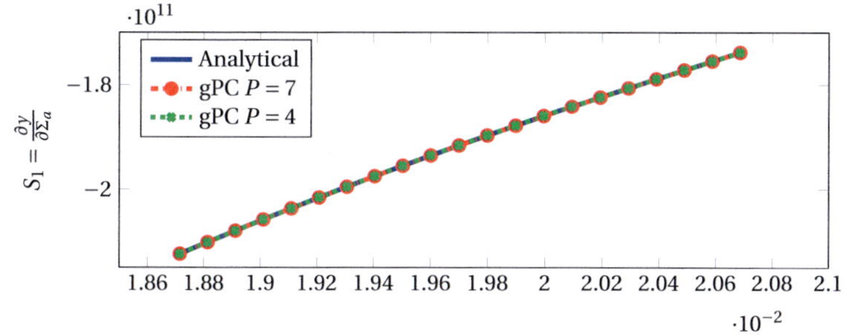

(a) Sensitivity of the parameter Σ_a by gPC approximation $P = 4$, $P = 7$, and analytical solution

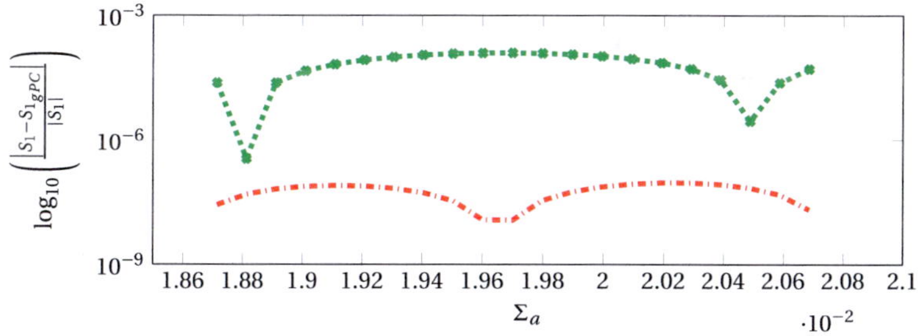

(b) Relative error of the approximation of the sensitivity by gPC in logarithmic scale

Figure 4.1: Comparison of the gPC approximation of the sensitivity of the parameter Σ_a with the analytic calculation

Figure 4.2 shows the relative error of the gPC approximation of the sensitivity of all parameters in logarithmic scale with respect to the the gPC order P. The upper plot of figure 4.2 presents the relative error of the gPC approximation of the sensitivities at the nominal value $\boldsymbol{q}^0 = [\Sigma_a^0, D^0, S^0, \Sigma_d^0]^T$. The relative error at tensorization points are evaluated and then averaged. The average relative error is presented in the lower plot. Both plots exhibit the convergence of the gPC approximation of the sensitivity. The approximations of the sensitivity of the parameter D have the largest relative error compared to the other parameters. This may be caused by the complexity of the analytical formula, which requires higher orders of the polynomial to imitate. However, the magnitude of the relative error of 10^{-4} for the polynomial of the order 4 should be sufficient for common applications.

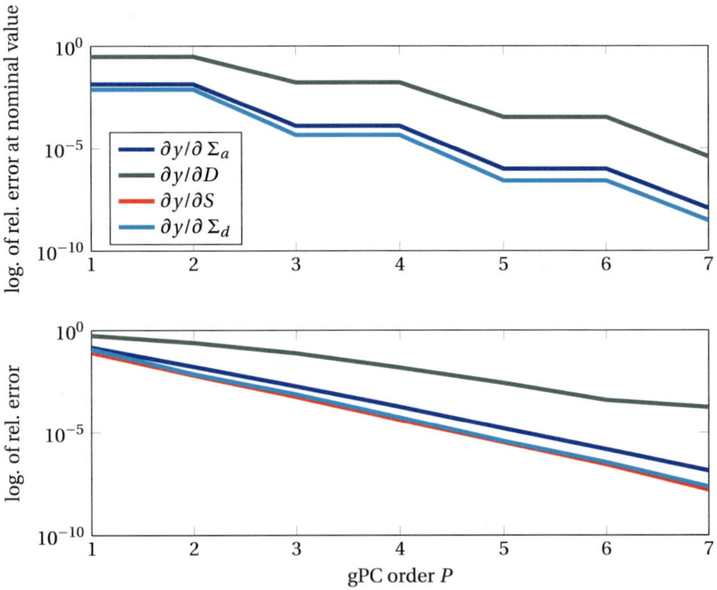

Figure 4.2: Error of the approximation of the derivative by the gPC with respect to the polynomial order P

In conclusion, this example shows that the local sensitivities approximated by the gPC give a satisfactory result. The error of the approximation depends on the order of the gPC and the complexity of the function. The main advantage of this method is that it considers the system as a black-box and does not need any analytical solution to calculate the derivative, which is usually not available for DPS. The information about the sensitivity of the system at various points in the parameter space is a useful information for the model user to understand the system behavior and can be used to optimize or improve the model.

4.3 Global sensitivity analysis

The *Global Sensitivity Analysis* (GSA) is the study of the global variation of a system response with regard to the input parameters. In contrast to the local sensitivity analysis, where inputs are varied about a nominal value, in global sensitivity analysis uncertainties caused by combinations of parameters throughout the permissible parameter space are examined.

Many global sensitivity analysis approaches have been proposed in the UQ community. Iooss reviews most of GSA methods in [Ioo15]. In the review, the author concludes the GSA methods in figure 4.3. Most of the GSA methods are categorized in two groups according to the paper, i.e. screening methods and variance-based methods. The screening methods focus on the identification of non-influential variables among a large number of input variables. In contrast, the variance-based methods are more precise and can provide quantitative relative importance of each parameter. The variance-based methods, therefore, require larger number of model evaluations compared to the screening methods, as indicated in figure 4.3. In the review, the authors locate the GSA methods based on their assumptions about the model complexity and regularity.

De Rocquigny [dR08] have proposed a decision tree flow chart (see figure 4.4) to help the practitioners choose the appropriate SA method for their problems. For our application, the relative importance of the model parameters is necessary to understanding the system behavior. Furthermore, the number of considered model parameters is not so large. According to these conditions, but without considering to the computational cost of the model evaluation, the Sobol method is the most appropriate method and also discussed explicitly in this thesis. The Sobol method is suitable for models with low computational cost. However, this computational cost issue can be solved with the generalized Polynomial Chaos expansion, which will be discussed next in this chapter. Readers interested in other global sensitivity analysis approaches are referred to [Sal04], [Ioo15] and references therein.

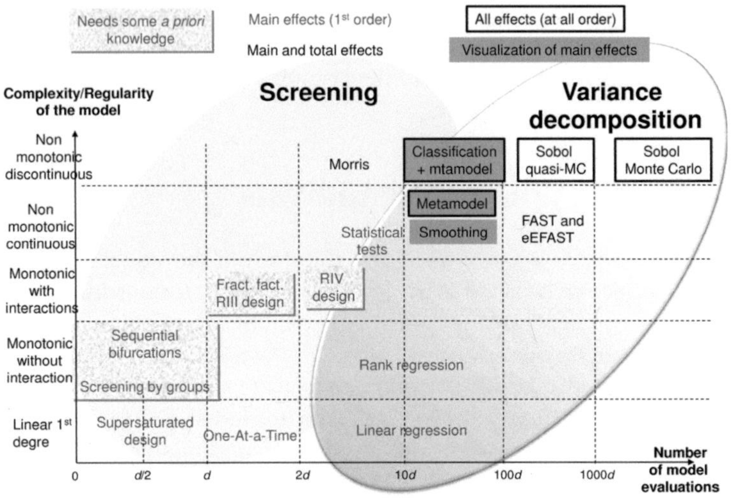

Figure 4.3: Classification of Global Sensitivity Analysis methods from [Ioo15]

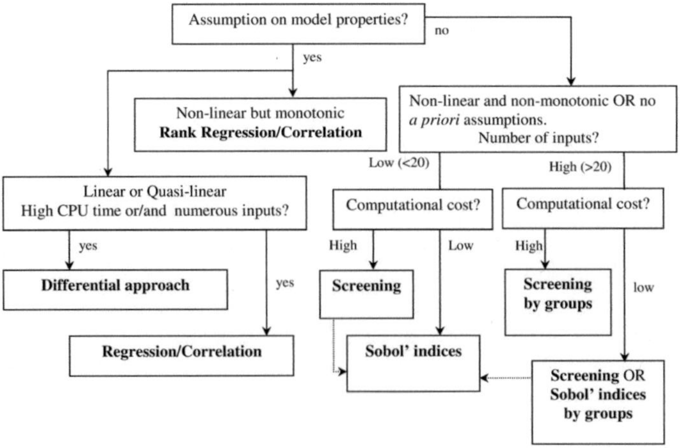

Figure 4.4: Decision diagram for choosing a GSA method from [dR08]

4.3.1 Sobol functional decomposition

The Sobol method is a variance-based global sensitivity analysis technique that has been applied to assess the relative importance of input parameters on the output. It results the Sobol sensitivity index as a normalized measure to determine the inputs importance. The method is based on the Sobol decomposition. The derivative of the Sobol decomposition discussed in the following is summarized from [Cre09], [Smi13] and [Sud08].

Considering the output y depending on the parameter vector $\boldsymbol{q}$, whose relation is described by some nonlinear model $y = f(\boldsymbol{q})$. The parameter vector $\boldsymbol{q}$ is considered to be uncertain and therefore modeled as a vector of random variables a known distribution. The random variable vector $\boldsymbol{Q}$ is composed of d independent and identically distributed RVs q_k with ranges Ω_k and pdfs $p_{Q_k}(q_k)$. The range for the random vector $\boldsymbol{Q}$ are then

$$\boldsymbol{\Omega} = \bigtimes_{k=1}^{d} \Omega_k. \tag{4.12}$$

Since the Q_k are independent, the joint density of $\boldsymbol{Q}$ is

$$p_{\boldsymbol{Q}}(\boldsymbol{q}) = \prod_{k=1}^{d} p_{Q_k}(q_k).$$

The distributions are assumed to be identical here only for the sake of the simplicity of notation. These relations can be extended to different distributions without difficulty.

The output Y is also a random variable and the nonlinear function $f(\cdot)$ is a mapping $\boldsymbol{q} \in \boldsymbol{\Omega} \to \mathcal{L}_2\left(\boldsymbol{\Omega}, p_{\boldsymbol{Q}}(\boldsymbol{q})\right)$. For every function $f \in \mathcal{L}_2\left(\boldsymbol{\Omega}, p_{\boldsymbol{Q}}(\boldsymbol{q})\right)$, the Sobol functional decomposition of f is then

$$f(\boldsymbol{q}) = \sum_{\mathcal{A} \subseteq \{1,2,\ldots,d\}} f_{\mathcal{A}}\left(\boldsymbol{q}_{\mathcal{A}}\right), \tag{4.13}$$

where $\mathcal{A} = \{i_1, \ldots, i_s\}$ is a set of integers with cardinality $s = \mathrm{card}(\mathcal{A})$, $\boldsymbol{q}_{\mathcal{A}} = \left[q_{i_1}, \ldots, q_{i_s}\right]^T$, and $f_\emptyset := f_0$. Each of the 2^d functions $f_{\mathcal{A}}$ of the decomposition, except for $f_\emptyset$, is assumed to satisfy

$$\int_{\Omega_k} f_{\mathcal{A}}\left(\boldsymbol{q}_{\mathcal{A}}\right) p_{Q_k}(q_k) dq_k = 0, \qquad \forall \mathcal{A} \ni k. \tag{4.14}$$

This implies the orthogonality of the function $f_{\mathcal{A}}$ in the following sense:

$$\int_{\mathbf{\Omega}} f_{\mathcal{A}}(\boldsymbol{q}_{\mathcal{A}})\, f_{\mathcal{B}}(\boldsymbol{q}_{\mathcal{B}})\, p_{\boldsymbol{Q}}(\boldsymbol{q}) d\boldsymbol{q} = 0, \qquad \forall \mathcal{A} \neq \mathcal{B}. \tag{4.15}$$

According to assumption (4.14), the Sobol decomposition is unique and the components of the function are computed by the relation:

$$f_{\mathcal{A}}(\boldsymbol{q}_{\mathcal{A}}) = \int_{\Omega^{d-s}} f(\boldsymbol{q}) p_Q(\boldsymbol{q}_{\sim\mathcal{A}}) d\boldsymbol{q}_{\sim\mathcal{A}} - \sum_{\substack{\mathcal{B}\subset\mathcal{A} \\ \mathcal{B}\neq\mathcal{A}}} f_{\mathcal{B}}(\boldsymbol{q}_{\mathcal{B}}), \tag{4.16}$$

where $\boldsymbol{q}_{\sim\mathcal{A}}$ denotes the vector having all the components of $\boldsymbol{q}$ except for those in the set $\mathcal{A}$, for example

$$\boldsymbol{q}_{\sim\{i\}} = [q_1, \ldots, q_{i-1}, q_{i+1}, \ldots, q_d]. \tag{4.17}$$

4.3.2 Sobol sensitivity indices

The response $Y = f(\boldsymbol{Q})$ has the variance σ^2 defined as

$$\sigma^2 := \mathrm{Var}(Y) = \int_{\mathbf{\Omega}} f^2(\boldsymbol{q}) p_{\boldsymbol{Q}}(\boldsymbol{q}) d\boldsymbol{q} - f_0^2. \tag{4.18}$$

Due to the orthogonality of the functions decomposed by the Sobol expansion, the variance can be expressed as

$$\sigma^2 = \sum_{\substack{\mathcal{A}\subset\{1,2,\ldots,d\} \\ \mathcal{A}\neq\emptyset}} \sigma^2_{\mathcal{A}}, \tag{4.19}$$

where $\sigma^2_{\mathcal{A}}$ is the partial variance defined as

$$\sigma^2_{\mathcal{A}} = \int_{\Omega^s} f^2_{\mathcal{A}}(\boldsymbol{q}_{\mathcal{A}}) p_{Q_{\mathcal{A}}}(\boldsymbol{q}_{\mathcal{A}}) d\boldsymbol{q}_{\mathcal{A}}.$$

$\sigma^2_{\mathcal{A}}$ can be expressed as a combination of conditional variances:

$$\sigma^2_{\mathcal{A}} = \mathrm{var}(\mathbb{E}(Y|\boldsymbol{q}_{\mathcal{A}})) - \sum_{\substack{\mathcal{B}\subset\mathcal{A} \\ \mathcal{B}\neq\mathcal{A},\mathcal{B}\neq\emptyset}} \sigma^2_{\mathcal{B}}. \tag{4.20}$$

The Sobol indices or variance-based sensitivity indicies are defined to be

$$S_{\mathcal{A}} := \frac{\sigma^2_{\mathcal{A}}}{\sigma^2}, \tag{4.21}$$

$$\text{so that} \quad \sum_{\substack{\mathcal{A}\subseteq\{1,2,\ldots,d\}\\ \mathcal{A}\neq\emptyset}} S_{\mathcal{A}} = 1. \tag{4.22}$$

These indices express the share of the variance of Y caused by of the given parameters. Each of Sobol sensitivity indices $S_{\mathcal{A}}$ is a sensitivity measure describing the variance due to the uncertainties in the set of the input parameters $\mathcal{A}$. The first-order indices S_i (card$(\mathcal{A}) = 1$) represent the influence of each parameter taken alone, whereas the higher-order indices (card$(\mathcal{A}) > 1$) give the sensitivity measures of the variance of Y due to the interaction between the parameters $\boldsymbol{q}_{\mathcal{A}}$, without taking into account the effect of the parameters $\boldsymbol{q}_{\mathcal{B}}$ for $\mathcal{B} \subset \mathcal{A}$ and $\mathcal{B} \neq \mathcal{A}$. For example the second-order sensitivity index $S_{i,j}$, indicates the sensitivity of Y due to the interaction between q_i and q_j, without taking into account the effect of each parameter separately, which are on the other hand measured by S_i and S_j.

To illustrate the Sobol method, the Sobol decomposition of a function of three RVs $Y = f(Q_1,Q_2,Q_3)$ is demonstrated as an example. According to equation (4.13), the Sobol expansion of the given function is:

$$\begin{aligned}
f(Q_1,Q_2,Q_3) = & f_0 && \text{(mean response)}\\
& + f_1(Q_1) + f_2(Q_2) + f_3(Q_3) && \text{(first-order)}\\
& + f_{12}(Q_1,Q_2) + f_{13}(Q_1,Q_3) + f_{23}(Q_2,Q_3) && \text{(second-order)}\\
& + f_{123}(Q_1,Q_2,Q_3). && \text{(third-order)}
\end{aligned}$$

The total variance σ^2 of the response Y can be expressed as:

$$\sigma^2 = \sigma^2_{\{1\}} + \sigma^2_{\{2\}} + \sigma^2_{\{3\}} + \sigma^2_{\{1,2\}} + \sigma^2_{\{1,3\}} + \sigma^2_{\{2,3\}} + \sigma^2_{\{1,2,3\}}.$$

where the partial variances are given by:

$$\sigma^2_{\{i\}} = \int_{\Omega_i} f_i^2(q_i) p_{Q_i}(q_i) dq_i \qquad \text{for first-order,}$$

$$\sigma^2_{\{i,j\}} = \int_{\Omega_i}\int_{\Omega_j} f_{ij}^2(q_i,q_j) p_{Q_j}(q_j) p_{Q_i}(q_j) dq_j dq_i \qquad \text{for second-order,}$$

$$\sigma^2_{\{1,2,3\}} = \int\limits_{\Omega_3}\int\limits_{\Omega_2}\int\limits_{\Omega_1} f^2_{123}(q_1,q_2,q_3)p_{\boldsymbol{Q}}(\boldsymbol{q})dq_3dq_2dq_1 \qquad \text{for third-order.}$$

The relation (4.22) in this case is:

$$S_{\{1\}} + S_{\{2\}} + S_{\{3\}} + S_{\{1,2\}} + S_{\{1,3\}} + S_{\{2,3\}} + S_{\{1,2,3\}} = 1.$$

The Sobol indices provide comprehensive measures for quantifying the influence of parameter uncertainty on the variance of the response. The number of indices grows, however, in an exponential way with dimension d, i.e. there are $2^d - 1$ indices altogether. For this reason, Homma and Satelli [Hom96] introduced the so-called *total indices* S_{T_i} to represent the *total effects* due to the parameter q_i. The total sensitivity indices are defined as:

$$S_{T_i} := \sum_{\mathcal{A}\ni i} S_{\mathcal{A}}. \tag{4.23}$$

For example, the total sensitivity index of the parameter q_2 for the response $Y = f(Q_1,Q_2,Q_3)$ is:

$$S_{T_2} = S_{\{2\}} + S_{\{1,2\}} + S_{\{2,3\}} + S_{\{1,2,3\}}. \tag{4.24}$$

For computational time and interpretation reasons, it is recommended to calculate the indices only for first two orders so that only the Sobol indices S_i and S_{ij} and total Sobol indices S_T have to be calculated in practice.

4.3.3 Computation of Sobol indices

Computing the Sobol indices (equation (4.21)) requires the computation of the variance σ^2 and the partial variances $\sigma^2_{\mathcal{A}}$. These variances can be estimated by sampling-based methods. Considering a sample set of N_{samp} realizations of the input variables $\{\boldsymbol{q}^i\}_{i=1}^{N_{\text{samp}}}$, the mean value f_0 and the variance σ^2 of the system response can be estimated from the sample as:

$$\hat{f}_0 = \frac{1}{N_{\text{samp}}} \sum_{i=1}^{N_{\text{samp}}} f(\boldsymbol{q}^i), \tag{4.25}$$

$$\hat{\sigma}^2 = \frac{1}{N_{\text{samp}}} \sum_{i=1}^{N_{\text{samp}}} f^2(\boldsymbol{q}^i) - \hat{f}_0^2. \tag{4.26}$$

To compute the partial variance $\sigma^2_{\mathcal{A}}$ from equation (4.3.2), the computation of the conditional variance $\mathrm{var}(\mathbb{E}(Y|\boldsymbol{q}_{\mathcal{A}}))$ is required. The conditional variance can be estimated from the sample by:

$$\begin{aligned}\mathrm{var}(\mathbb{E}(Y|\boldsymbol{q}_{\mathcal{A}})) &= \mathbb{E}\left(\mathbb{E}\left(Y|\boldsymbol{q}_{\mathcal{A}}\right)^2\right) - \mathbb{E}\left(\mathbb{E}\left(Y|\boldsymbol{q}_{\mathcal{A}}\right)\right)^2 \\ &= \mathbb{E}\left(\mathbb{E}\left(Y|\boldsymbol{q}_{\mathcal{A}}\right)^2\right) - \mathbb{E}\left(Y\right)^2 \\ &\approx \frac{1}{N_{\mathrm{samp}}}\sum_{i=1}^{N_{\mathrm{samp}}}\left(\frac{1}{N_{\mathrm{samp}}}\sum_{j=1}^{N_{\mathrm{samp}}} f(\boldsymbol{q}^j_{\sim\mathcal{A}}, \boldsymbol{q}^i_{\mathcal{A}})\right)^2 - \hat{f}_0^2. \end{aligned} \tag{4.27}$$

The computational cost of the estimation of the conditional variance by equation (4.27) is of $\mathcal{O}(N^2_{\mathrm{samp}})$, which is too expensive in practice. Sobol [Sob90] proposed a less expensive approach to approximate the Sobol indices by Monte-Carlo sampling. This is achieved by using two independent sample sets $\{\boldsymbol{q}^i\}_{i=1}^{N_{\mathrm{samp}}}$ and $\{\boldsymbol{\xi}^i\}_{i=1}^{N_{\mathrm{samp}}}$. The estimation of the conditional variance from the samples is:

$$\hat{\sigma}^2_{\mathcal{A}} = \frac{1}{N_{\mathrm{samp}}}\sum_{i=1}^{N_{\mathrm{samp}}} f(\boldsymbol{q}^i) f(\boldsymbol{\zeta}^i_{\mathcal{A}}) - \sum_{\substack{\mathcal{B}\subset\mathcal{A}\\ \mathcal{B}\neq\mathcal{A}}} \hat{\sigma}^2_{\mathcal{B}}, \tag{4.28}$$

where $(\zeta_j)^i_{\mathcal{A}} = \begin{cases} q^i_j, & \text{for } j \in \mathcal{A} \\ \xi^i_j, & \text{otherwise} \end{cases}.$

Finally, the estimation of the Sobol indices from the samples is obtained by

$$\hat{S}_{\mathcal{A}} = \frac{\hat{\sigma}^2_{\mathcal{A}}}{\hat{\sigma}^2}. \tag{4.29}$$

The samples used in this method can be achieved by more efficient methods such as LHS technique or QMC. The reader is referred to [Sob90] and [Cre09] for more elaborate mathematical formulations of this estimation. By using this method the computational cost of the conditional variance is reduced in $\mathcal{O}(2N_{\mathrm{samp}})$ [Cre09]. However, the computation of these indices are still expensive in case of an extensive computational model, since it still needs a large number of evaluations of the model in the parameter space.

This is the reason, why the Sobol method is not suitable for extensive computational models. One solution of this problem is using a surrogate model instead of the full model to estimate the variance. Another option is approximating the variance with the generalized Polynomial Chaos expansion as will be discussed in the next section.

4.4 Global sensitivity analysis with gPC

B. Sudret and T. Crestaux have independently shown the relationship between the Sobol indices and the gPC coefficients in [Sud08] and [Cre09] respectively. From equation (3.5), for $f \in \mathcal{L}_2(\boldsymbol{\Omega}, p_{\boldsymbol{Q}}(\boldsymbol{q}))$, the function can be approximated using the PCE truncated at order P:

$$Y = f(Q_1,\dots,Q_d) \approx a_0\Xi_0 + \sum_{i_1=1}^{d} a_{i_1}\Xi_1\left(Q_{i_1}\right) + \sum_{i_1=1}^{d}\sum_{i_2=1}^{i_1} a_{i_1 i_2}\Xi_2\left(Q_{i_1},Q_{i_2}\right) + \sum_{i_1=1}^{d}\sum_{i_2=1}^{i_1}\sum_{i_3=1}^{i_2} a_{i_1 i_2 i_3}\Xi_3\left(Q_{i_1},Q_{i_2},Q_{i_3}\right) + \dots. \tag{4.30}$$

One can see that, this PCE (4.30) is similar to the Sobol decomposition. For example, the function in the example $Y = f(Q_1,Q_2,Q_3)$ can be expanded with the polynomial chaos:

$$\begin{aligned} f(Q_1,Q_2,Q_3) \approx\ & a_0\Xi_0 + a_1\Xi_1(Q_1) + a_2\Xi_1(Q_2) + a_3\Xi_1(Q_3) + \\ & a_{1,1}\Xi_2\left(Q_1^2\right) + a_{2,2}\Xi_2\left(Q_2^2\right) + a_{3,3}\Xi_2\left(Q_3^2\right) + \\ & a_{1,2}\Xi_2(Q_1,Q_2) + a_{1,3}\Xi_2(Q_1,Q_3) + a_{2,3}\Xi_2(Q_2,Q_3) + \\ & a_{1,1,1}\Xi_3\left(Q_1^3\right) + a_{2,2,2}\Xi_2\left(Q_2^3\right) + a_{3,3,3}\Xi_2\left(Q_3^3\right) + \\ & a_{1,1,2}\Xi_3\left(Q_1^2,Q_2\right) + a_{1,1,3}\Xi_3\left(Q_1^2,Q_3\right) + \\ & a_{1,2,2}\Xi_3\left(Q_1,Q_2^2\right) + a_{2,2,3}\Xi_3\left(Q_2^2,Q_3\right) + \\ & a_{1,3,3}\Xi_3\left(Q_1,Q_3^2\right) + a_{2,3,3}\Xi_3\left(Q_2,Q_3^2\right) + \\ & a_{1,2,3}\Xi_3(Q_1,Q_2,Q_3) + \dots. \end{aligned} \tag{4.31}$$

The Sobol functions can be approximated by the PCE for example

$$\begin{aligned} f_1(Q_1) &\approx a_1\Xi_1(Q_1) + a_{1,1}\Xi_2\left(Q_1^2\right) + a_{1,1,1}\Xi_3\left(Q_1^3\right) + \dots \\ f_{12}(Q_1,Q_2) &\approx a_{1,2}\Xi_2(Q_1,Q_2) + a_{1,1,2}\Xi_3\left(Q_1^2,Q_2\right) + a_{1,2,2}\Xi_3\left(Q_1,Q_2^2\right) + \dots \\ f_{123}(Q_1,Q_2,Q_3) &\approx a_{1,2,3}\Xi_3(Q_1,Q_2,Q_3) + \dots. \end{aligned} \tag{4.32}$$

The PCE can be rewritten by using the multi-index notation $\underline{i} = (i_1, i_2, \ldots, i_d)$ with $|\underline{i}| = \sum_{\alpha=1}^{d} i_\alpha$

$$Y = \sum_{|\underline{i}|=0}^{N_P} \beta_{\underline{i}} \Psi_{\underline{i}}(Q_1, Q_2, \ldots, Q_d). \tag{4.33}$$

The element $f_{\mathcal{A}}$ of the Sobol decomposition f is approximated by

$$f_{\mathcal{A}} \approx \sum_{\underline{i} \in \underline{i}_{\mathcal{A}}} \beta_{\underline{i}} \Psi_{\underline{i}}(Q_1, Q_2, \ldots, Q_d), \tag{4.34}$$

where $\underline{i}_{\mathcal{A}} := \{\underline{\alpha} \in \underline{i} | \alpha_k > 0, \forall k = 1, \ldots, d\,, k \in \mathcal{A} \wedge \alpha_k = 0, \forall k = 1, \ldots, d\,, k \notin \mathcal{A}\}$. It is noted that this formulation can be generalized to the gPCE for other types of pdfs by using the others orthogonal polynomials according to table 3.1 as basis functions. As shown in section 3.1, the total variance can be calculated directly from the gPC coefficients and the basis functions:

$$\hat{\sigma}^2 = \sum_{|\underline{i}|=1}^{N_P} \beta_{\underline{i}}^2 \left\| \Psi_{\underline{i}}^2 \right\|. \tag{4.35}$$

Due to the orthogonality of the gPC basis, the conditional variance $\sigma_{\mathcal{A}}^2$ is:

$$\hat{\sigma}_{\mathcal{A}}^2 = \sum_{\underline{i} \in \underline{i}_{\mathcal{A}}} \beta_{\underline{i}}^2 \left\| \Psi_{\underline{i}}^2 \right\|. \tag{4.36}$$

Finally, the gPC-approximated Sobol indicies are:

$$\hat{S}_{\mathcal{A}} = \frac{\sum_{\underline{i} \in \underline{i}_{\mathcal{A}}} \beta_{\underline{i}}^2 \left\| \Psi_{\underline{i}}^2 \right\|}{\sum_{|\underline{i}|=1}^{N_P} \beta_{\underline{i}}^2 \left\| \Psi_{\underline{i}}^2 \right\|} \tag{4.37}$$

Moreover, analogous to equation (4.23) the gPC-approximated total Sobol indicies can be calculated by:

$$\hat{S}_{T_i} = \sum_{\mathcal{A} \ni i} \hat{S}_{\mathcal{A}}. \tag{4.38}$$

4.4.1 Performance evaluation of global sensitivity analysis computation using gPC by means of a numerical example

To illustrate the efficiency of the gPC method, the global sensitivity analysis is performed to the neutron diffusion example [Cac14a] (cf. section 3.4 and 4.2.1). The uncertainties of the parameters are assumed to be independent normal distribution as also assumed in the paper, which is equivalent to the example in section 3.4. The pdf of the RV is a normal distribution as:

$$\boldsymbol{Q} \sim \mathcal{N}\left(\boldsymbol{q}^0, \mathbf{C}_{\boldsymbol{Q}}\right),$$
$$\boldsymbol{q}^0 = \begin{bmatrix} 0.0197 & 0.16 & 10^7 & 7.438 \end{bmatrix}^T,$$
$$\mathbf{C}_{\boldsymbol{Q}} = \begin{bmatrix} (9.85 \times 10^{-5})^2 & 0 & 0 & 0 \\ 0 & (8.00 \times 10^{-3})^2 & 0 & 0 \\ 0 & 0 & (1.50 \times 10^{6})^2 & 0 \\ 0 & 0 & 0 & (7.44 \times 10^{-1})^2 \end{bmatrix}.$$

Using the gPC approximation constructed in the example in 3.4, the Sobol indices are computed from the gPC coefficients as shown in equation (4.37). All of the gPC computations are accomplished by the software for the gPC frameworks established for this thesis (see section 3.3). The Sobol indices of the system response at a distance $x = 10$ cm regarding to the different polynomials order of the gPC are presented in the table 4.1. The GSA results by the gPC approximation are compared with the sampling-based method described in section 4.3.3. The computation of Sobol indices by the sampling-based method is achieved by the Bourinet's code *FERUM* ver.4.1 [Bou10]. The results for the Sobol indices by the sampling-based method are also presented in table 4.1.

For this example, the exact values of all Sobol indices are not known. But the Sobol indices by the gPC with the polynomial order $P = 8$ agrees with the indices by the Quasi-Monte-Carlo with 10^7 samples, which gives a good estimation for the Sobol indices in this example. The Sobol indices by crude Monte-Carlo with $N = 10^7$ differ a little bit from the other two approaches. That is to be expected, as the MC has a slower convergence rate compared to the QMC.

Method	S_{Σ_a} in %	S_D in %	S_S in %	S_{Σ_d} in %	$S_{\{HO\}}$ in %	Number of model evaluations
gPC $P = 2$	7.275	$\approx 0^*$	64.185	28.531	0.009	9
gPC $P = 4$	7.119	$\approx 0^*$	63.104	28.046	1.731	49
gPC $P = 6$	7.165	$\approx 0^*$	63.515	28.229	1.091	201
gPC $P = 8$	7.181	$\approx 0^*$	63.656	28.291	0.872	681
MC $N = 10^3$	1.796	3.039	45.687	32.061	1.258	10^3
MC $N = 10^5$	8.097	0.028	63.168	27.567	1.141	10^5
MC $N = 10^7$	7.235	$\approx 0^*$	63.337	28.425	1.003	10^7
QMC $N = 10^3$	8.233	2.135	62.413	26.769	0.450	10^3
QMC $N = 10^5$	7.324	$\approx 0^*$	63.059	28.621	0.996	10^5
QMC $N = 10^7$	7.179	$\approx 0^*$	63.691	28.240	0.890	10^7

Table 4.1: Sobol indices approximated by the generalized polynomial chaos (gPC), Monte-Carlo (MC) and Quasi-Monte-Carlo (QMC) for each parameter of the system response of the neutron diffusion model with the required number of model evaluation, $\approx 0^*$ means approximately zero due to the same order of numerical artifacts. $S_{\{HO\}}$ is summation of Sobol indices higher than first-order

Table 4.1 shows that the results by the sampling-based methods with a small number of samples, e.g. $N = 10^3$, cannot provide a satisfactory result. The sampling-based method requires a large number of model evaluations in order to achieve a good result. In contrast, the gPC approximation requires only a small number of model evaluations as shown in table 4.1. The best case of the gPC method, gPC with the polynomial order $P = 8$, requires less than a thousand model callings. The worst gPC approximation with $P = 2$ needs only 9 model callings, and can provide an approximation comparable to the QMC with $N = 10^5$. This shows that the gPC is able to provide a very good approximation by only a small number of the model evaluations. This performance of the gPC method is very advantageous in case of extensive computational models e.g. for FEM.

The gPC software developed in this thesis offers a visualization of the sensitivity information, e.g. the pie chart shown in figure 4.5. This figure illustrates to which account the uncertainty of the considered system response is caused by the uncertainties of the parameters S, Σ_d and Σ_a respectively.

Comparing these Sobol indices with the local sensitivity from the example in section 4.2.1, one can suspect some contradiction at the first glance. The LSA gives the following sensitivities at the nominal values:

$$\frac{\partial y}{\partial \Sigma_a} = -1.92 \times 10^{11}, \qquad \frac{\partial y}{\partial D} = -1.33 \times 10^{5},$$
$$\frac{\partial y}{\partial S} = 3.77 \times 10^{2}, \qquad \frac{\partial y}{\partial \Sigma_d} = 5.08 \times 10^{8}.$$

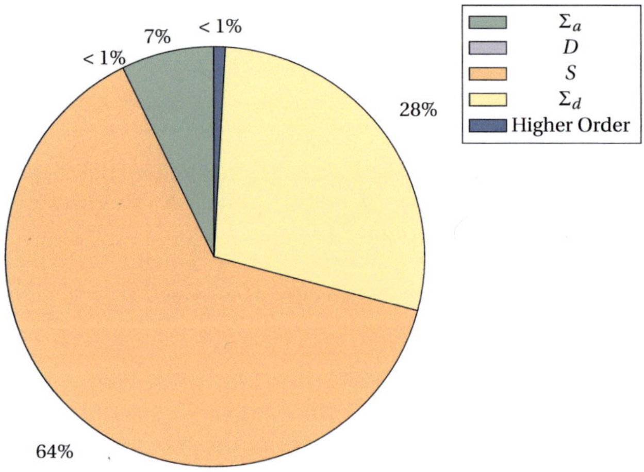

Figure 4.5: Visualization of the Sobol indices of the scalar system response at $x = 10$ cm

The sensitivities of both methods are actually not contradictory, but they state two different kind of information about the system response. The LSA provide the sensitivity at the nominal value without considering the range of the parameter space. The system response at the nominal value $\boldsymbol{q}^0$ is mostly sensitive to the change of the parameter Σ_a, and then to the parameters Σ_d, D and S respectively.

In contrast to the LSA, in the GSA the admissible parameter space is involved in the evaluation of the global sensitivities. With respect to the standard deviations of the parameters in this example, the parameter S is the most uncertain parameter with $\Delta S^0/S^0 = 15\%$. The parameter Σ_d has the standard deviation value of $\Delta\Sigma_d^0/\Sigma_d^0 = 10\%$, while the parameters Σ_a and D have relative standard deviations of 5%. This fact is also consistent with the Sobol indices results.

Most of the studies about GSA proposed in the aforementioned literature and publications work with scalar functions. In the framework developed for this thesis, the GSA can be applied to the DPS. This can be achieved by discretization of the infinite-dimensional field to a finite-dimensional field, which can be described by a random vector (see sections 2.2.1 and 3.1.2). In the implementation of the developed software, it means that the vector of gPC coefficients $[\beta_1,\ldots,\beta_{N_P}]$ is extended to matrix $[\boldsymbol{\beta}_1,\ldots,\boldsymbol{\beta}_{N_P}]$, while all other parts of the algorithm remain the same.

The result of applying GSA to DPS is visualized in figure 4.6. The figure presents the Sobol indices for all discretized positions along the x-axis. The figure gives information about the influence of the parameters in the whole domain. The Sobol indices at the position of $x = 10$ cm (dashed line) is equivalent to the result of figure 4.5. It can be seen that in the middle (from $x = -40$ cm to $x = 40$ cm) the sensitivity remains similar. At the boundaries, the parameter Σ_a has less influence and the parameter D gains a little influence on the system responses.

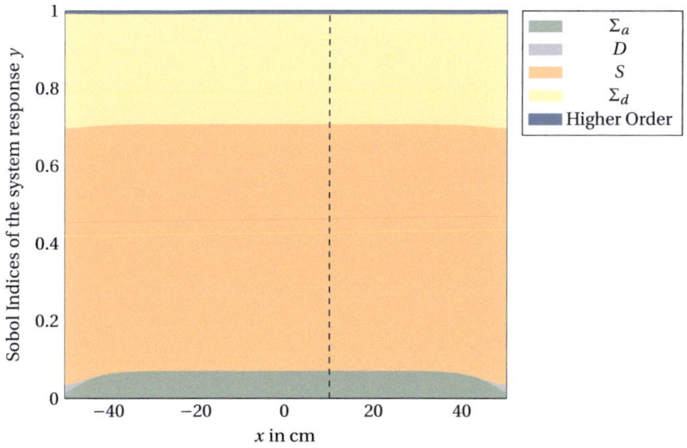

Figure 4.6: Global Sensitivity Analysis of the response of the neutron diffusion system

5

A Bayesian Approach to Parameter Calibration

In the developed framework, all imperfections of a model and incomplete information are encompassed in form of uncertainties of parameters. All various imperfections in the modeling lead to a lack of agreement between the solution of physics-based models and the observations from the real processes. On the assumption that the model is perfect, the solution of the error-free computational model will depend solely on the model parameters. In order that the model predictions correspond to the real measurement, the uncertain parameters should be calibrated with the observed data. The ASME guide [ame06] gives the definition of model calibration as

> *"The process of adjusting physical modeling parameters in the computational model to improve agreement with experimental data."*

The calibration is typically required in the modeling of complex physical processes. Parameter calibration can be considered as part of the broader field of parameter estimation, which refers to procedures for estimating any type of parameters in a model using supplied data. Due to the wide range in which this kind of procedures can get in the field of modeling and simulation, Oberkampf [Obe10] has classified the procedure in three different activities regarding to the meaningfulness of the parameters. These are:

- Parameter measurement: Determination of physically meaningful parameters, that can be independently measured.
- Parameter estimation: Determination of physically meaningful parameters, that cannot be independently measured in practice.
- Parameter calibration: Adjustment of parameters that have little or no physical meaning.

Although there is a difference between the terms estimation and calibration, as the calibrated parameter does not necessarily represent physical reality, both problems deal with the same question: Which values of the parameters contributing to the computational model will yield the prediction closest to the given data?

The parameter estimation or respectively calibration problem can be formulated as an inverse problem. Solving an inverse problem is the process of calculating the causal factors from a set of observations. For the parameter estimation, it means the determination of parameters in computational models from the observations. Inverse problems are typically ill-posed regarding the the definition given by Jacques Hadamard [Had23].

An inverse problem can be solved deterministically or statistically. The deterministic solving can be achieved by defining some measure of discrepancy between the model prediction and the observation. Then, the discrepancy is minimized by some optimization algorithms. Due of the nature of ill-posed problems, the minimum of such a problem is often not unique. An appropriate regularization is required, in order to obtain the unique solution of this problem. Some short overview of the deterministic methods can be found in [Kai05].

In contrast, the statistical approach reformulates the inverse problem as an inference problem, where the regularization of the problem is expressed in form of a prior distribution. The deterministic approach results in a single point in the parameter space as a solution of the optimization problem, compared to the statistical approach, which provides more information about the estimated parameters in from of a posterior probability distribution function. The theory of statistical inverse problems can be found in [Kai05, Tar05].

The Bayesian statistic formulation for identification and parameter calibration problem is more appropriate, as the framework of this thesis works under probabilistic UQ approach. However, the Bayesian approach requires excessive computation in general, and some techniques are needed to facilitate the computation. The Polynomial Chaos Expansion (PCE) can be applied to reduce the computational effort.

The Bayesian parameter calibration approach with the help of gPCE is integrated to the developed comprehensive framework of analyzing and identification. In this thesis, the gPCE is applied to the parameter calibration in two ways, which belong to the thesis contributions. The first one is using the gPC approximation as the surrogate model similarly to application to uncertainty propagation and sensitivity analysis mentioned in previous chapters. The second fashion is expanding the recursive Bayesian estimator with the polynomial chaos basis. This technique is quite new, provides good results and has attractive properties.

The chapter is organized in the following. The Bayesian statistic formulation of an inverse problem and the typical batch determination of the posterior pdf are discussed firstly. Then, the Bayesian updating with PCE is proposed. The usage of the gPC as a surrogate model is discussed next. Finally, all approaches discussed in this chapter are demonstrated by means of a numerical example.

5.1 Bayesian statistic formulation of inverse problem

From a statistical point of view, the inverse problem is reformulated as a problem of statistical inference by means of Bayesian statistics [Kai05, Tar05]. The Bayesian inference is broadly used to extract the information from the measured data in many fields, as reviewed in [vT11]. Using Bayesian inference to calibrate the parameters of a computational model can be found in many publications, e.g. [Hig12, Ken00]. Especially, applying Bayesian parameter estimation to distributed parameter systems can be found in e.g. [Wan04], [Lie10] and [HD13].

In this section, the Bayesian approaches to parameter calibration in distributed parameter systems (DPSs) are summarized from all the mentioned publications and are integrated into the developed framework.

Given a DPS described with equation (2.4), the solution operator $\mathcal{M}(\cdot)$ of equation (2.4) describes the explicit relationship between the spatial distributed state $x(\boldsymbol{r},t)$ of the system and model parameter $\boldsymbol{q}$ as:

$$x(\boldsymbol{r},t) = \mathcal{M}\left(\boldsymbol{q}\right) \tag{5.1}$$

For computational purposes, the distributed quantities such as the state $x(\boldsymbol{r},t)$ or the source $s(\boldsymbol{r},t)$ are usually approximated into a finite space with discretization by means of a set of spatial basis function $\{\phi_i\}_{i=1}^{N_s}$ as showed in section 2.2.1. The corresponding coefficients of the basis functions are encompassed in the vector form e.g. $\boldsymbol{x}(t) = \left[x_1(t),\ldots,x_{N_s}(t)\right]^T$, or $\mathbf{s}(t) = \left[s_1(t),\ldots,s_{N_s}(t)\right]^T$.

As discussed in the section 2.1.3, all parameters in DPS are then collected in the parameter vector $\boldsymbol{q}$. An example of the statistical formulation of inverse problem of PDE can be found in the previous study [Ja12]. Using the observation operator (2.5) the noise-free forward model of the DPS is defined as:

$$\boldsymbol{y}^M(\boldsymbol{q}) = \mathcal{M}_m\left(x(\boldsymbol{r},t),\boldsymbol{w}=0\right) \tag{5.2}$$

In the course of this thesis, additive white Gaussian noise is assumed, which covers most typical applications in the practice. The observed value $\boldsymbol{y}$ is determined as:

$$\boldsymbol{y} = \boldsymbol{y}^M(\boldsymbol{q}) + \boldsymbol{w}, \tag{5.3}$$

where $\boldsymbol{w}$ is the vector of the realization from the Gaussian distributed random vector with zero mean and the covariance $\mathbf{C}_W$. In the statistical inverse problem, the parameters $\boldsymbol{q}$ to be estimated are modeled as a random vector $\boldsymbol{Q}(\omega)$. The randomness of the unknown parameters describes the degree of information concerning their realizations in Bayesian statistics (see chapter 2). The degree of information concerning there values is coded in the form of probability distributions. The primary solution of a statistical inverse problem is the posterior distribution. This solution in form of a probability distribution is the main difference between the statistical approach and deterministic approach, which gives only the optimal single point solution. By a statistical inverse problem a single point solution can be derived from the probability distribution by using expectation operator $\mathbb{E}(\cdot)$.

The ill-posedness of an inverse problem is handled by restating the problem as a well-posed extension in a large space of probability distributions. Moreover, it is allowed to add a priori knowledge that is often hidden within the deterministic regularization view. This *prior information* of unknown parameters $\boldsymbol{q}$ can be coded into a probability density. This probability density called *a priori* pdf $p_{\mathrm{pr}}(\boldsymbol{q})$ represents the *priori* knowledge of the modeler about the parameters.

With a joint probability density function of the unknown parameters $\boldsymbol{q}$ and the measurement $\boldsymbol{y}$, which is denoted by $p(\boldsymbol{q},\boldsymbol{y})$, there is the relation :

$$p(\boldsymbol{q},\boldsymbol{y}) = p(\boldsymbol{y}|\boldsymbol{q})p(\boldsymbol{q}) = p(\boldsymbol{q}|\boldsymbol{y})p(\boldsymbol{y}), \tag{5.4}$$

where $p(\boldsymbol{y}|\boldsymbol{q})$ is the conditional pdf of $\boldsymbol{Y}$ given the parameter value $\boldsymbol{q}$. By given the observation $\boldsymbol{y}$, the $p(\boldsymbol{y}|\boldsymbol{q})$ is considered as a function of $\boldsymbol{q}$ and is called a *likelihood function* $L(\boldsymbol{q})$. The conditional probability $p(\boldsymbol{q}|\boldsymbol{y})$, called *posterior* pdf of $\boldsymbol{q}$, expresses the information about $\boldsymbol{q}$ after having received an observation $\boldsymbol{y}$.

In the Bayesian framework, the inverse problem is expressed in following way: Given the measurement data $\boldsymbol{y}$, find the conditional probability distribution $p(\boldsymbol{q}|\boldsymbol{y})$ of the variable $\boldsymbol{q}$. Using the Bayes' rule, which is well-known for more than two centuries, the statistical formulation of the inverse problem can be concluded in the following theorem [Kai05]:

Theorem 1 (Bayes' theorem of inverse problems) *Assume that the random variable $\mathbf{Q}(\omega) \in \mathbb{R}^d$ has a known prior probability density $p_{pr}(\boldsymbol{q})$ and the data consist of the observed value $\mathbf{y}$ of an observable random variable $\boldsymbol{Y}(\omega) \in \mathbb{R}^n$ such that $p(\mathbf{y}) > 0$. Then the posterior probability distribution of $\boldsymbol{q}$, given the data $\mathbf{y}$ is*

$$p_{post}(\boldsymbol{q}) = p(\boldsymbol{q}|\boldsymbol{y}) = \frac{p(\boldsymbol{y}|\boldsymbol{q})p_{pr}(\boldsymbol{q})}{p(\boldsymbol{y})} \tag{5.5}$$

□

The marginal probability density in equation (5.5)

$$p(\boldsymbol{y}) = \int_{\mathbb{R}^d} p(\boldsymbol{q},\boldsymbol{y})d\boldsymbol{q} = \int_{\mathbb{R}^d} p(\boldsymbol{y}|\boldsymbol{q})p_{\mathrm{pr}}(\boldsymbol{q})d\boldsymbol{q} \tag{5.6}$$

acts only as a normalization constant and therefore is neglected in general.

Comparing to the deterministic approach, specifying the likelihood function is related to determination of the measure of the discrepancy between the model prediction and the observation. The prior distribution, which is required in order that the problem is well-posed, is associated with the deterministic regularization. The advantage of the statistic approach to the deterministic method is the flexibility to specify the measure and the regularization, which can be done by determining the likelihood function and the prior pdf. The Bayesian approach also offers a comprehensible formulation of the pdf from the regularity assumption.

In summary, solving the inverse problem from the Bayesian point of view can be divided into three subtasks.

1. Determine an appropriate prior probability density $p_{\text{pr}}(\boldsymbol{q})$ relied on the available prior information of unknown $\boldsymbol{q}$.

2. Construct an appropriate likelihood function $p(y|\boldsymbol{q})$ that describes the interrelation between the observation and the unknown.

3. Compute the posterior probability density $p_{\text{post}}(\boldsymbol{q})$.

Firstly, the prior density can be modeled as discussed in section 2.1.5. Commonly it is modeled by a Gaussian density, which can be formulated in following form.

$$p_{\text{pr}}(\boldsymbol{Q}) = \left(\frac{1}{2\pi \det(\mathbf{C}_{\boldsymbol{Q}})}\right)^{d/2} \exp\left(-\frac{1}{2}(\boldsymbol{q}-\boldsymbol{q}_0)^T \mathbf{C}_{\boldsymbol{Q}}^{-1}(\boldsymbol{q}-\boldsymbol{q}_0)\right), \tag{5.7}$$

where $\mathbf{C}_{\boldsymbol{Q}} \in \mathbb{R}^{d\times d}$ denotes the covariance matrix and $\boldsymbol{q}_0$ the means of the random variables $\boldsymbol{Q}(\omega)$.

The likelihood function $L(\boldsymbol{q})$ is specified regarding to the underlying assumptions about the distribution of errors. Under the assumption of additive Gaussian noise in the measurement and given the forward model $\boldsymbol{y}^M(\boldsymbol{q})$ the likelihood function is:

$$L(\boldsymbol{q}) = \left(\frac{1}{2\pi \det(\mathbf{C}_{\boldsymbol{W}})}\right)^{n/2} \exp\left(-\frac{1}{2}(\boldsymbol{y}^M(\boldsymbol{q})-\boldsymbol{y})^T \mathbf{C}_{\boldsymbol{W}}^{-1}(\boldsymbol{y}^M(\boldsymbol{q})-\boldsymbol{y})\right), \tag{5.8}$$

where $\mathbf{C}_{\boldsymbol{W}}$ denotes the covariance of the measurement noise. For the construction of likelihoods for other error models, the reader may refer to [Kai05].

Finally, the method to compute the posterior distribution $p_{\text{post}}(\boldsymbol{q})$ has to be determined. The evaluation of the posterior pdf $p_{\text{post}}(\boldsymbol{q})$ requires an integration in general. However, the analytic calculation of the integral is feasible only for some trivial problems. Therefore, the posterior pdf is often estimated by statistically based sampling methods as e.g. the Monte-Carlo (MC) method. Some fundamental sampling algorithms for MC are e.g. *Rejection Sampling, Importance Sampling, Sampling Importance*

Resampling. In Bayesian Statistics, one often uses the Markov-Chain-Monte-Carlo (MCMC) [Gel14] technique. The MCMC method uses a set of points from the given distribution, a *sample*, to approximate the Monte-Carlo integration. The sample ensembles are generated by using the Markov chain random walk algorithm, for example *Metropolis-Hasting Algorithm* or *Gibbs Sampling.* There are also some modifications of this algorithm for better performance and computation such as *Delay Rejection (DR)* or *Adaptive Metropolis-Hasting (AM)* [Haa06]. The MCMC realizes the computation of the posterior distribution and its estimation of expected values. The reader is referred to [Gel14, Haa06] for theory and implementation of MCMC and its extension.

However, the MCMC approach requires a large number of repeated solving of the forward model leading to an extensive computational effort especially in case of solving partial differential equations (PDEs). Therefore, two new approaches using Polynomial Chaos to solve this problem are proposed, which is one of the contributions of this thesis.

5.2 Recursive Bayesian estimation (Bayesian updating)

Another approach to evaluate the posterior distribution belongs to the so-called *recursive Bayesian estimation* methods, which estimates the unknown pdf recursively over time using incoming measurements and a mathematical model. At the beginning of the estimation all the information about the parameter $\boldsymbol{q}$ is contained by a prior distribution $p_{\mathrm{pr}}(\boldsymbol{q})$. The measurements are assumed to be conditionally independent, obtained one at time and arranged in a sequence $\mathbf{Y}_n = [\boldsymbol{y}_1, \ldots, \boldsymbol{y}_n]$ according the time. The recursive Bayesian estimation begins by evaluating the posterior pdf after the first measurement with the prior distribution

$$p(\boldsymbol{q}|\boldsymbol{y}_1) = \frac{p(\boldsymbol{y}_1|\boldsymbol{q})p_{\mathrm{pr}}(\boldsymbol{q})}{p(\boldsymbol{y}_1)}. \tag{5.9}$$

Bayes' theorem can be formulated in the recursive form by using the posterior pdf from the previous time step (t_{n-1}) as the prior pdf at the current time step (t_n). The pdf of the parameter $\boldsymbol{q}$ is estimated every time step by integrating the sequence of sensor measurements $\mathbf{Y}_n = [\boldsymbol{y}_1, \ldots, \boldsymbol{y}_n]$. The posterior distribution at time step n has the form:

$$p(\boldsymbol{q}|\mathbf{Y}_n) = \frac{p(\boldsymbol{y}_n|\boldsymbol{q})p(\boldsymbol{q}|\mathbf{Y}_{n-1})}{p(\boldsymbol{y}_n|\mathbf{Y}_{n-1})}. \tag{5.10}$$

The Bayesian inference in each time step after obtaining the new measurement $\boldsymbol{y}_k$ is known as *Bayesian Updating*. Because a reordering of the measurements does not affect the final solution, the posterior pdf at the final time step is exactly the posterior pdf achieved by the batch method.

Due to the sequential nature of the recursive Bayesian estimation, it is broadly applied to dynamic systems to estimate the state $\boldsymbol{x}(t)$, which can change in time. The extension of the Bayesian estimation for a dynamic system is known as *Sequential Bayesian Filtering* or also known as *Bayesian Filter*. Two assumptions are required to derive the sequential Bayesian filtering:

- The states follow a 1st-order Markov process $p(\boldsymbol{x}_n|\boldsymbol{x}_{0:n-1}) = p(\boldsymbol{x}_n|\boldsymbol{x}_{n-1})$.
- The observation at the k−th time step depends only upon the current state, so is conditionally independent of all other states given the current states $p(\boldsymbol{y}_k|\boldsymbol{x}_{0:k}) = p(\boldsymbol{y}_n|x_k)$.

The concept of Sequential Bayesian Filtering is broadly used in control and robotics. The term is sometimes more specified as:

- *Filtering*: When the pdf of the current state $\boldsymbol{x}_k$ is estimated by given past and current observations $\mathbf{Y}_k = [\boldsymbol{y}_1, \ldots, \boldsymbol{y}_k]$:

$$p(\boldsymbol{x}_k|\mathbf{Y}_k). \tag{5.11}$$

- *Smoothing*: When the pdf of a past state $\boldsymbol{x}_k$ is estimated by given past and present observations $\mathbf{Y}_n = [\boldsymbol{y}_1, \ldots, \boldsymbol{y}_n]$:

$$p(\boldsymbol{x}_k|\mathbf{Y}_n), \text{ for } k < n. \tag{5.12}$$

- *Prediction*: When a future state $\boldsymbol{x}_k$ is estimated by given past and current observations $\mathbf{Y}_n = [\boldsymbol{y}_n, \ldots, \boldsymbol{y}_n]$:

$$p(\boldsymbol{x}_k|\mathbf{Y}_n), \text{ for } k > n. \tag{5.13}$$

For the parameter calibration in this thesis, the parameter vector $\boldsymbol{q}$ is considered to be time invariant. Therefore, there is no distinction between these three terms.

One of the most well-known methods of sequential Bayesian filtering is the *Kalman-Filter* (KF). The Kalman-Filter [Kal60] can be seen as the special case of the sequential Bayesian filtering. It is an optimal recursive estimator of the state $\boldsymbol{x}$, in the sense that it is unbiased and is a minimum-variance estimator. However the KF is limited only for linear systems with the multivariate normal distributions.

Because most systems behave nonlinearly, and the pdf cannot remain Gaussian, the Kalman-Filter is advanced in two directions, namely linearization of the nonlinear model and approximation the non-Gaussian distribution. The linearization the nonlinear model with the Taylor series results the *Extended Kalman Filter* (EKF) see e.g. [Jaz07, Gre11]. The EKF can be applied to nonlinear systems, however, it can estimate only a Gaussian distribution.

The *Unscented Kalman Filter* (UKF) [Jul04] uses a deterministic sampling technique known as the unscented transform to approximate the mean and the covariance of the distribution from the sigma points.

Another approach is the approximation of non-Gaussian distribution based on Monte-Carlo. This sequential Monte-Carlo based are for example the *Particle Filter* or *Ensemble Kalman Filter* (EnKF). The Particle Filter and the EnKF are not limited to Gaussian distributions but based on sampling, which requires extensive computational effort. More theory and details about the Bayesian filtering can be found in many publications, e.g. [Che03, Thr05, Gre11, Sär13].

5.2.1 Linear Bayesian updating with PCE

As mentioned before, that an approximation for nonlinear or non-gaussian distributions without sampling is needed. Recently, Hermann G. Matthies et al. proposed a numerical strategy by applying functional approximation methods to recursive Bayesian estimation, which can estimate non-Gaussian distributions without sampling in [Paj13, Paj12, Ros12b, Ros13]. They formulate an abstract linear filter from the relation between the Gauss-Markov theorem and the Kalman-Filter [Lue97], and named it as **Gauss-Markov-Kalman-Filter** (GMKF)[Mat16b]. The reader is referred to [Mat16b] for the elaborated mathematical derivation of the filter.

In this thesis, the idea of the GMKF is applied especially to the parameter calibration problem by using the PCE as the functional approximation methods. The estimation problem of the parameter vector $\boldsymbol{q}$ from the measurement $\boldsymbol{y}$ is established as described in the section 5.1. The error-free model prediction of the mathematical model $\boldsymbol{y}^M(\boldsymbol{q})$ from equation (5.2) is related to the measurement regarding to equation (5.14).

The linear minimum-variance Bayesian updating when additional data becomes available is obtained by the orthogonal projection according to the following theorem paraphrased from [Paj12]:

Theorem 2 *Assume that some uncertain knowledge about the parameter $\boldsymbol{q}$ is available, so that the parameter $\boldsymbol{q}$ can be considered as RV $\boldsymbol{Q}^f(\omega)$ (with a superscript f denoting "forecast"* [1]*) in a Bayesian fashion. When the measurement $\boldsymbol{Y}(\omega)$ becomes available regarding the assumption of the additive noise*

$$\boldsymbol{Y}(\omega) = \boldsymbol{Y}^M(\boldsymbol{Q}^f(\omega)) + \boldsymbol{W}(\omega), \tag{5.14}$$

the orthogonal projection $\boldsymbol{Q}^a(\omega)$ (with a superscript a denoting "assimilated" [2]*) on the subspace spanned by $\boldsymbol{Q}^f(\omega)$ and $\boldsymbol{Y}(\omega)$ is the best estimator of $\boldsymbol{q}$ in the $\mathcal{L}_2$-norm.*

$$\boldsymbol{Q}^a(\omega) = \boldsymbol{Q}^f(\omega) + \mathbf{K}\big(\boldsymbol{Y}(\omega) - \boldsymbol{Y}^M(\omega)\big) \tag{5.15}$$

with the Kalman Gain operator $\boldsymbol{K}$

$$\mathbf{K} := \mathbf{C}_{\boldsymbol{Q}^f \boldsymbol{Y}^M}\left(\mathbf{C}_{\boldsymbol{W}} + \mathbf{C}_{\boldsymbol{Y}^M}\right)^{-1}, \tag{5.16}$$

where $\mathbf{C}_W$ is the covariance of the noise and the other covariances are given by:

$$\mathbf{C}_{\boldsymbol{Q}^f \boldsymbol{Y}^M} = \mathbb{E}\left[(\boldsymbol{Q}^f - \overline{\boldsymbol{Q}}^f) \otimes (\boldsymbol{Y}^M - \overline{\boldsymbol{Y}}^M)\right], \tag{5.17}$$

$$\mathbf{C}_{\boldsymbol{Y}^M} = \mathbb{E}\left[(\boldsymbol{Y}^M - \overline{\boldsymbol{Y}}^M) \otimes (\boldsymbol{Y}^M - \overline{\boldsymbol{Y}}^M)\right], \tag{5.18}$$

where $\otimes$ is the tensor product (see definition (11) *in Appendix A).* □

---

1 Instead of "forecast", the term "predict" is also used in some literature.

2 The terms "update" or "analysis" are used instead of "assimilated"in some literature.

This theorem can be seen as a generalization of the well-known Gauss-Markov theorem [Lue97].

According to the works in [Paj13, Paj12, Ros12b, Ros13, Mat16b], the RVs $\boldsymbol{Q}(\omega), \boldsymbol{Y}(\omega), \boldsymbol{Y}^{\boldsymbol{M}}(\omega)$ can be represented by using PCE as the series of orthogonal polynomials Ψ_k.

$$\boldsymbol{Q}(\omega) = \sum_{k=0}^{N_P} \boldsymbol{q}_k \Psi_k(\boldsymbol{\xi}(\omega)) \tag{5.19}$$

$$\boldsymbol{Y}(\omega) = \sum_{k=0}^{N_P} \boldsymbol{y}_k \Psi_k(\boldsymbol{\xi}(\omega)) \tag{5.20}$$

$$\boldsymbol{Y}^M(\omega) = \sum_{k=0}^{N_P} \boldsymbol{y}_k^M \Psi_k(\boldsymbol{\xi}(\omega)) \tag{5.21}$$

According to the central limit theorem, the noise is assumed to be Gaussian. Therefore, the multi-dimensional Hermite polynomials $\boldsymbol{H}_k(\boldsymbol{\xi}(\omega))$are applied as the PC basis. Using this PCE representation, the equation (5.15) can be expanded to

$$\sum_{k=0}^{N_P} \boldsymbol{q}_k^a \boldsymbol{H}_k(\boldsymbol{\xi}(\omega)) = \sum_{k=0}^{N_P} \boldsymbol{q}_k^f \boldsymbol{H}_k(\boldsymbol{\xi}(\omega)) + \mathbf{K}\left(\sum_{k=0}^{N_P} \boldsymbol{y}_k \boldsymbol{H}_k(\boldsymbol{\xi}(\omega)) - \sum_{k=0}^{N_P} \boldsymbol{y}_k^M \boldsymbol{H}_k(\boldsymbol{\xi}(\omega))\right) \tag{5.22}$$

Using the orthogonality property of the PC basis from the equation (3.13), one may take the inner product of the Eq.(5.22) with the PC basis $\boldsymbol{H}_k$ according to Eq.(3.35). This can be done by multiplying the Eq.(5.22) with each $\boldsymbol{H}_k$ and taking the expectation of the obtained equation. Further all obtained equations for all k are divided it by $\|\boldsymbol{H}_k^2\|$ so that

$$\boldsymbol{q}_k^a = \boldsymbol{q}_k^f + \mathbf{K}(\boldsymbol{y}_k - \boldsymbol{y}_k^M), \qquad k = 0, \ldots, N_P. \tag{5.23}$$

The column vectors $\boldsymbol{q}_k^f$ are arranged into a matrix $\mathbf{Q}^f$

$$\mathbf{Q}^f := \begin{bmatrix} \mathbf{q}_0^f & \cdots & \mathbf{q}_k^f & \cdots & \mathbf{q}_{N_P}^f \end{bmatrix} \tag{5.24}$$

and so on for $\boldsymbol{q}_k^a, \boldsymbol{y}_k, \boldsymbol{y}_k^M$. That leads to the equation in the matrix form:

$$\mathbf{Q}^a = \mathbf{Q}^f + \mathbf{K}(\mathbf{Y} - \mathbf{Y}^M). \tag{5.25}$$

The covariance matrices $\mathbf{C}_{Q^f Y^M}, \mathbf{C}_{Y^M}$ can be determined from the PCE directly with the polynomial chaos algebra as mentioned in the section 3.1.2. For the random vector represented in the PCE $\boldsymbol{X} = \sum_k^{N_P} \boldsymbol{x}_k \boldsymbol{H}_k(\boldsymbol{\xi}(\omega))$, its covariance can be written in the matrix form as:

$$\mathbf{C}_X = \check{\mathbf{X}}\mathbf{G}\check{\mathbf{X}}^T. \tag{5.26}$$

$\check{\mathbf{X}}$ is a matrix containing the PC coefficients of the random vector without the $k = 0$ term (the mean) and $\mathbf{G} = (g_{kl})$ is the diagonal Gram matrix with $g_{kl} = \mathbb{E}[\Psi_k(\boldsymbol{\xi}(\omega))\Psi_l(\boldsymbol{\xi}(\omega))]$. In case of Hermite polynomials, $g_{kl} = \mathbb{E}[\boldsymbol{H}_k(\boldsymbol{\xi}(\omega))\boldsymbol{H}_l(\boldsymbol{\xi}(\omega))] = \delta_{kl}k!$, this also lead to the following equation:

$$\mathbf{C}_X = \sum_{k,l>0} x_k \otimes x_l k! \tag{5.27}$$

The linear Bayesian update procedure with the PCE can be concluded in and implemented with the following algorithm.

Algorithm of Linear Bayesian update with PCE

1. Approximation a priori information $\boldsymbol{Q}^0(\omega)$ by PCE as described in the chapter 3 and applied as the forecast variable $\boldsymbol{Q}^f(\omega)$, $\mathbf{Q}^f := \begin{bmatrix}\mathbf{q}_0^f & \cdots & \mathbf{q}_k^f & \cdots & \mathbf{q}_{N_P}^f\end{bmatrix}$ and then centralize (take out the mean) to $\tilde{\mathbf{Q}}^f := \begin{bmatrix}\mathbf{q}_1^f & \cdots & \mathbf{q}_k^f & \cdots & \mathbf{q}_{N_P}^f\end{bmatrix}$.

2. Represent the available measurement $\boldsymbol{Y}(\omega)$ by PCE $\mathbf{Y} := \begin{bmatrix}\mathbf{y}_0 & \cdots & \mathbf{y}_k & \cdots & \mathbf{y}_{N_P}\end{bmatrix}$ and then centralize to $\tilde{\mathbf{Y}} := \begin{bmatrix}\mathbf{y}_1 & \cdots & \mathbf{y}_k & \cdots & \mathbf{y}_{N_P}\end{bmatrix}$.

3. Solve the stochastic forward problem and forecast measurement $\boldsymbol{Y}^M(\omega)$ based on the current $\boldsymbol{Q}^f(\omega)$.

4. Represent the $\boldsymbol{Y}^M(\omega)$ in the PCE form $\mathbf{Y}^M := \begin{bmatrix}\mathbf{y}_0^M & \cdots & \mathbf{y}_k^M & \cdots & \mathbf{y}_{N_P}^M\end{bmatrix}$ then centralize to $\tilde{\mathbf{Y}}^M := \begin{bmatrix}\mathbf{y}_1^M & \cdots & \mathbf{y}_k^M & \cdots & \mathbf{y}_{N_P}^M\end{bmatrix}$.

5. Compute the covariances $\mathbf{C}_W$, $\mathbf{C}_{Y^M}$ and $\mathbf{C}_{Q^f Y^M}$.

$$\mathbf{C}_W = \check{\mathbf{Y}}\mathbf{G}\left(\check{\mathbf{Y}}\right)^T \tag{5.28}$$

$$\mathbf{C}_{Y^M} = \check{\mathbf{Y}}^M\mathbf{G}\left(\check{\mathbf{Y}}^M\right)^T \tag{5.29}$$

$$\mathbf{C}_{Q^f Y^M} = \check{\mathbf{Q}}^f\mathbf{G}\left(\check{\mathbf{Y}}^M\right)^T \tag{5.30}$$

6. Compute the Kalman Gain $\mathbf{K}$ according to equation (5.16).

7. The assimilated variable in PCE form $\mathbf{Q}^a$ is computed according to the update equation (5.46).

8. Using the assimilated variable $\mathbf{Q}^a$ of the current step as the forecast variable $\mathbf{Q}^f$ for the next step and then repeat the procedure of step 3, in case that the measurement does not change. Go to step 2, if a new measurement is available.

Kalman-Filter as a special case

It is shown in the [Paj13] that the Kalman-Filter can be considered as a special case of this linear Bayesian update with the PCE. The original Kalman-Filter has two equations in the update step:

$$\boldsymbol{x}_{k|k} = \boldsymbol{x}_{k|k-1} + \mathbf{K}_k\left(\boldsymbol{y}_k - \boldsymbol{y}^M(\boldsymbol{x}_{k|k})\right) \qquad \text{update mean} \tag{5.31}$$

$$\mathbf{C}_{\boldsymbol{x}_{k|k}} = (\mathbf{I} - \mathbf{K}_k\mathbf{L}_k)\,\mathbf{C}_{\boldsymbol{x}_{k|k-1}} \qquad \text{update covariance} \tag{5.32}$$

with the measurement model

$$\boldsymbol{y}^M(\boldsymbol{x}_{k|k}) = \mathbf{L}_k\boldsymbol{x}_{k|k}. \tag{5.33}$$

By setting $\boldsymbol{x}_{k|k} = \boldsymbol{q}^a$ and $\boldsymbol{x}_{k|k-1} = \boldsymbol{q}^f$ in equation (5.23), the linear Bayesian update with PCE for $k = 0$ provides exact the same equation as the update mean equation of the original Kalman-Filter.
The covariance update equation can be obtained by computing the covariance of $\boldsymbol{q}^a$. From equation (5.34) it yields:

$$\mathbf{C}_{Q^a} = \tilde{\mathbf{Q}}^a\mathbf{G}\left(\tilde{\mathbf{Q}}^a\right)^T = \left(\tilde{\mathbf{Q}}^f + \mathbf{K}\left(\tilde{\mathbf{Y}} - \tilde{\mathbf{Y}}^M\right)\right)\mathbf{G}\left(\tilde{\mathbf{Q}}^f + \mathbf{K}\left(\tilde{\mathbf{Y}} - \tilde{\mathbf{Y}}^M\right)\right)^T \tag{5.34}$$

By assuming that the measurement errors are not correlated with the state forecast, i.e. $\mathbf{C}_{\boldsymbol{Q}^a\boldsymbol{W}} = 0$, one obtains

$$\mathbf{C}_{\boldsymbol{Q}^a} = \mathbf{C}_{\boldsymbol{Q}^f} + \mathbf{K}\mathbf{C}_W\mathbf{K}^T + \mathbf{K}\mathbf{C}_{Y^M}\mathbf{K}^T - \mathbf{C}_{\boldsymbol{Q}^f\boldsymbol{Y}^M}\mathbf{K}^T - \mathbf{K}\mathbf{C}^T_{\boldsymbol{Q}^f\boldsymbol{Y}^M} \tag{5.35}$$

$$= \mathbf{C}_{\boldsymbol{Q}^f} + \mathbf{K}\left(\mathbf{C}_W + \mathbf{C}_{\boldsymbol{Y}^M}\right)\mathbf{K}^T - \mathbf{C}_{\boldsymbol{Q}^f\boldsymbol{Y}^M}\mathbf{K}^T - \mathbf{K}\mathbf{C}^T_{\boldsymbol{Q}^f\boldsymbol{Y}^M} \tag{5.36}$$

Inserting **K** from Eq. (5.16) results

$$\mathbf{C}_{\boldsymbol{Q}^a} = \mathbf{C}_{\boldsymbol{Q}^f} + \mathbf{C}_{\boldsymbol{Q}^f\boldsymbol{Y}^M}\left(\mathbf{C}_{\boldsymbol{W}} + \mathbf{C}_{\boldsymbol{Y}^M}\right)^{-1}\mathbf{C}^T_{\boldsymbol{Q}^f\boldsymbol{Y}^M} - 2\mathbf{C}_{\boldsymbol{Q}^f\boldsymbol{Y}^M}\left(\mathbf{C}_{\boldsymbol{W}} + \mathbf{C}_{\boldsymbol{Y}^M}\right)^{-1}\mathbf{C}^T_{\boldsymbol{Q}^f\boldsymbol{Y}^M} \tag{5.37}$$

$$= \mathbf{C}_{\boldsymbol{Q}^f} + \mathbf{C}_{\boldsymbol{Q}^f\boldsymbol{Y}^M}\left(\mathbf{C}_{\boldsymbol{W}} - \mathbf{C}_{\boldsymbol{Y}^M}\right)^{-1}\mathbf{C}^T_{\boldsymbol{Q}^f\boldsymbol{Y}^M} \tag{5.38}$$

By setting $\boldsymbol{x}_{k|k} = \boldsymbol{q}^a$ and $\boldsymbol{x}_{k|k-1} = \boldsymbol{q}^f$ in Eq. (5.37) and $\mathbf{C}^T_{\boldsymbol{x}_{k|k-1}Y^M} = \mathbf{L}_k\mathbf{C}_{\boldsymbol{x}_{k|k-1}}$ for the linear measurement model (5.33), one obtains exactly the covariance update equation (5.32). This shows that the KF can be considered as the special case of the linear Bayesian updating with PCE. The Kalman-Filter provides the first two statistical moments, whereas the linear Bayesian update can provide higher order statistical moments. The information of the higher order moments is in the higher order PCE polynomial terms. The estimated distribution can be approximated by PCE. Therefore, this approach can be applied to nonlinear systems with non-Gaussian distribution.

5.3 gPCE as surrogate model

The main difficulty of solving the Bayesian inference in inverse problems is the extensive computational effort. In order to compute the posterior distribution, the forward model in the likelihood function has to be solved many times. Several approaches to overcome this obstacle have been proposed in literature lately. Some studies apply *Model Order Reduction* (MOR) to the forward model to reduce the computation e.g. [Lie10, Wan04]. Another approach is using surrogate models such as *Gaussian process emulation* in [Hig10, Ken00]. As shown in chapter 3, the gPC approximation can be used as a surrogate model. Marzouk applied the gPC approximation to the Bayesian inverse problem in [Mar09]. The posterior distribution can be approximated by using the gPC in the evaluation of the likelihood function. It also has been proven in the paper that the approximated posterior distribution converges to the exact posterior w.r.t. the Kullback-Leibler divergence (KLD) measure.

The gPC surrogate can be considered as a polynomial surface response model of parameters. By considering the parameters with their specified ranges as independently uniformly distributed RVs, the model can be approximated by

$$\boldsymbol{y}^M \approx \hat{\boldsymbol{y}}^M(\boldsymbol{q}) = \sum_{k=0}^{N_P} \boldsymbol{\beta}_k \boldsymbol{\Psi}_k(\boldsymbol{q}), \tag{5.39}$$

where gPC basis functions Ψ_k are constructed from 1D-Legendre polynomials $Le_k(\cdot)$ due to the uniform distributions of RVs. In this thesis the application of gPC as surrogate model to the parameter calibration problem is proposed.

Then the gPC approximation is used as the forward model (5.2) in the evaluation of the likelihood function (5.8) as:

$$L(\boldsymbol{q}) = \left(\frac{1}{2\pi \det(\mathbf{C}_W)}\right)^{n/2} \exp\left(-\frac{1}{2}(\hat{\boldsymbol{y}}^M(\boldsymbol{q}) - \boldsymbol{y})^T \mathbf{C}_W^{-1} (\hat{\boldsymbol{y}}^M(\boldsymbol{q}) - \boldsymbol{y})\right) \tag{5.40}$$

This formulation can be applied to both batch and recursive methods to approximate the posterior distribution. This approach is very useful in case that the original model $\boldsymbol{y}^M$ requires an extensive computation. The limitation of this method is that the gPC approximation is valid only in the defined parameter space. The sampling outside the defined parameter space could lead to erroneous result. To avoid that problem, one should use the defined parameter space as a prior of the inverse problem, or a truncated Gaussian distribution [Xiu04b] in case of a prior normal distribution.

5.4 Performance evaluation of parameter calibration using gPC by means of a numerical example

To demonstrate the parameter calibration method, the neutron diffusion problem from [Cac14a] is taken as an example (cf. examples in section 3.4, 4.2.1 and 4.4.1). The neutron diffusion system and the uncertain parameters are described in section 3.4. The uncertain parameters should be calibrated by giving the experimental measurement of $y = 3.40 \times 10^9$ neutrons $\text{cm}^{-3}\text{s}^{-1}$ with a relative standard deviation of 5% as shown in figure 5.1.

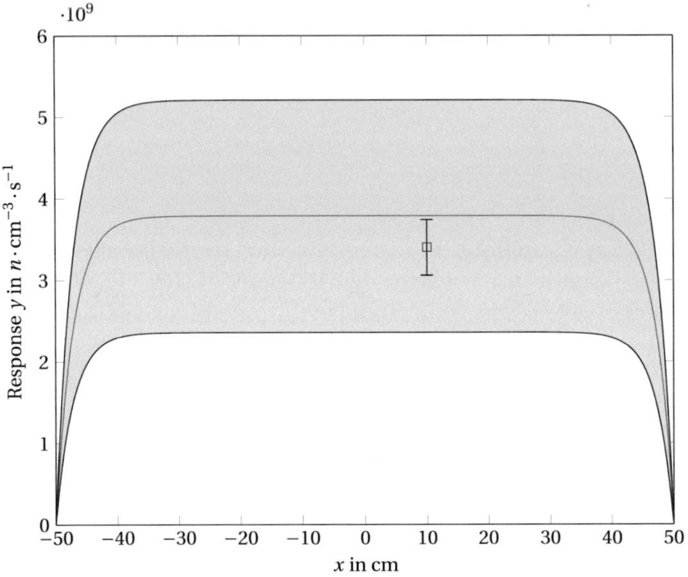

Figure 5.1: Measurement and its standard deviation in the neutron diffusion example denoted by □ and the indicated interval. The red solid line and the gray area present the model response prediction based on the prior distribution of the parameters before calibration.

Cacuci et al. proposed the framework called *Best-Estimate Model Calibration* (BEST-EST) [Bad12, Ars14] for assimilation of experimental data and computational information. The method calibrates the model parameters, which encompass in parameter vector $\boldsymbol{q}$, by using the sensitivity matrix $\mathbf{J}$ (see Eq. (4.2)). For notation consistency, w.r.t. to the given observation $\boldsymbol{y}$, the calibrated parameter vector by the BEST-EST method $\boldsymbol{q}^{\text{be}}$ of the system response $\boldsymbol{y}^M(\boldsymbol{q}^0)$ can be calculated by

$$\boldsymbol{q}^{\text{be}} = \boldsymbol{q}^0 + \underbrace{\left(\mathbf{C}_{\boldsymbol{Q}\boldsymbol{Y}^M} + \mathbf{C}_{\boldsymbol{Q}}\mathbf{J}^T\right)\left(\mathbf{C}_W + \mathbf{C}_{\boldsymbol{Y}^M\boldsymbol{Q}}\mathbf{J}^T + \mathbf{J}\mathbf{C}_{\boldsymbol{Q}\boldsymbol{Y}^M} + \mathbf{C}_{\boldsymbol{Y}^M}\right)^{-1}}_{\text{Kalman Gain } \mathbf{K}}\left(\boldsymbol{y} - \boldsymbol{y}^M\right). \tag{5.41}$$

Although the Kalman-Filter and the Kalman gain are not mentioned in the papers, equation (5.41) is similar to the linear Bayesian update (5.15). Due to the Jacobi matrix **J**, this BEST-EST approach can be considered as the Extended Kalman smoothing with correlated noises. He applied this method in [Cac14a] to calibrate the parameters which results in mean values and the covariance matrix of the calibrated parameters. The covariance matrix of the calibrated parameters is decomposed into the standard deviation of each parameter and the correlation matrix for comparison reasons. The mean values, standard deviation and correlation matrix resulted in [Cac14a] are summarized in the tables 5.1, 5.2 and 5.3 respectively.

In this example, the Bayesian inverse formulation described in this chapter is preformed to calibrate the parameters. As mentioned previously, there are two main approaches namely the batch and the recursive methods. In the batch method, a posterior distribution of the parameters is assessed by a MCMC method. The other approach shown in this example is the linear Bayesian updating with PCE (see section 5.2.1). The gPC approximation is also used as surrogate model to show the feasibility of the method.

Under the Bayesian inverse formulation, one needs a prior distribution. For all methods discussed in this example, the uncertainty of the parameters is expressed in the prior distribution:

$$Q \sim \mathcal{N}\left(\boldsymbol{q}^0, \mathbf{C}_Q\right)$$

$$\boldsymbol{q}^0 = \begin{bmatrix} 0.0197 & 0.16 & 10^7 & 7.438 \end{bmatrix}^T$$

$$\mathbf{C}_Q = \begin{bmatrix} (9.85\times10^{-5})^2 & 0 & 0 & 0 \\ 0 & (8.0\times10^{-3})^2 & 0 & 0 \\ 0 & 0 & (1.5\times10^{6})^2 & 0 \\ 0 & 0 & 0 & (7.44\times10^{-1})^2 \end{bmatrix}$$

Batch method

Following the formulation of the Bayesian inverse problem as described in section 5.1, in this subsection, the full model is used in the likelihood function (Eq.(5.8)). It should be noted that this is feasible because of the availability of the analytical solution. The computation of a posterior in the presented framework is based on the MCMC code of Haario [Haa06]. The burn-in with $N_{\text{burn-in}} = 1000$ iterations is firstly done to find an appropriate starting point. In this example, the MCMC is run with a chain length of

$N_{\text{MCMC}} = 10000$ iterations. The samples of the posterior distribution are obtained as a result. This result is visualized as approximated marginal pdfs in figure 5.2 and scatter plots in figure 5.3. The statistical values, namely the mean values, the standard deviation and the correlation matrix, are calculated from the samples and summarized in the tables 5.1, 5.2 and 5.3 respectively.

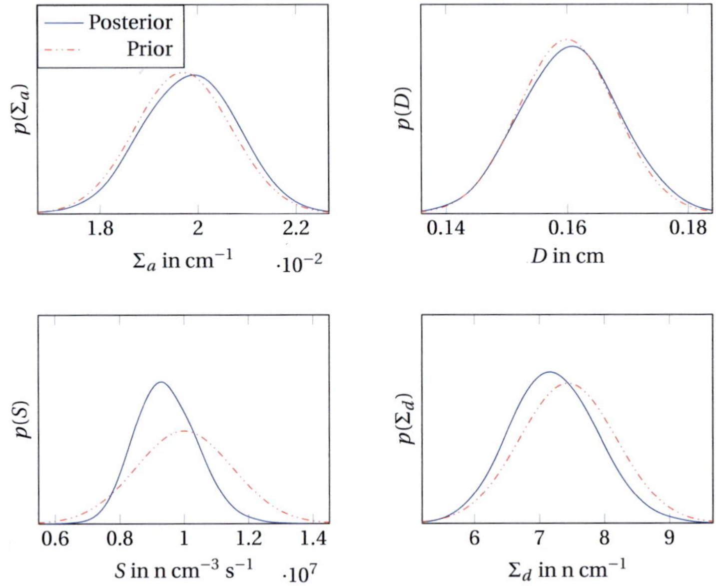

Figure 5.2: Prior pdf and posterior pdf of the calibrated parameter by the batch method for the diffusion neutron problem

Bayesian update with PCE

The setting of the Bayesian update with PCE is formulated as described in section 5.2.1. The parameter vector $\boldsymbol{q}$, model responses $\boldsymbol{y}^M$ and the measurements $\boldsymbol{y}$ are represented by PCE and summarized in matrix form:

$$\mathbf{Q} = \left[\mathbf{q}_0 \cdots \mathbf{q}_k \cdots \mathbf{q}_{N_P}\right], \tag{5.42}$$

$$\mathbf{Y} = \left[\mathbf{y}_0 \cdots \mathbf{y}_k \cdots \mathbf{y}_{N_P}\right], \tag{5.43}$$

$$\mathbf{Y}^M = \left[\mathbf{y}_0^M \cdots \mathbf{y}_k^M \cdots \mathbf{y}_{N_P}^M,\right]. \tag{5.44}$$

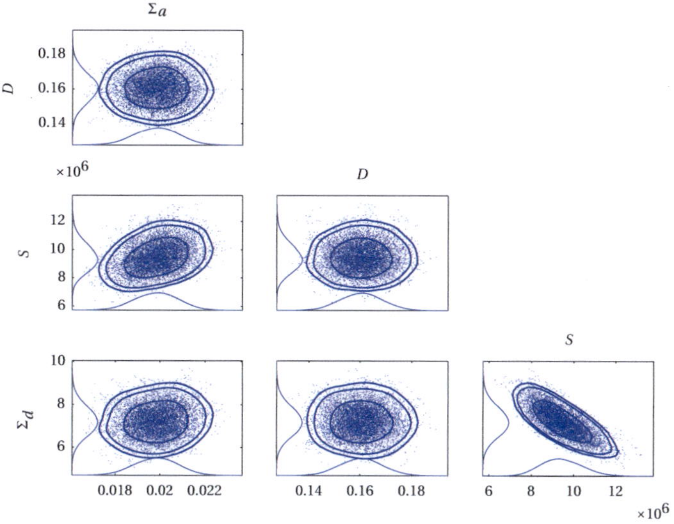

Figure 5.3: Scatter plots of the samples from the posterior distribution assessed by batch method

In this example, the PCE of order $P = 4$ is used and the dimension d of the parameters is 4. Using Eq. (3.12), it yields

$$N_P + 1 = \frac{(P+d)!}{P!d!} = \frac{(4+4)!}{4!4!} = 70.$$

There are totally 70 coefficients for each parameter with $N_P = 69$ in this case. For example, the prior distribution of the parameter $\boldsymbol{q}$ can be rewritten in the matrix form as

$$\mathbf{Q}^0 = \begin{bmatrix} \mu_{\Sigma_a} & \sigma_{\Sigma_a} & 0 & 0 & 0 & 0 & \cdots \\ \mu_D & 0 & \sigma_D & 0 & 0 & 0 & \cdots \\ \mu_S & 0 & 0 & \sigma_S & 0 & 0 & \cdots \\ \mu_{\Sigma_d} & 0 & 0 & 0 & \sigma_{\Sigma_d} & 0 & \cdots \end{bmatrix} \tag{5.45}$$

The assimilation of measurements and model response is achieved by applying theorem 2. The algorithm described in section 5.2.1 is implemented in the developed software. The PC coefficients of matrix $\mathbf{Q}$ are updated according to equation (5.46) as

$$\mathbf{Q}^j = \mathbf{Q}^{j-1} + \mathbf{K}\left(\mathbf{Y} - \mathbf{Y}^M\right), \qquad \text{for } j = 1, \ldots, N_{\text{step}}. \tag{5.46}$$

In this example, the parameters are updated in $N_{\text{step}} = 5$ steps to demonstrate the method. The mean values and the 2σ intervals of each step are illustrated in figure 5.4. The mean and the standard deviation are calculated from the PCE coefficients of $\mathbf{Q}$.

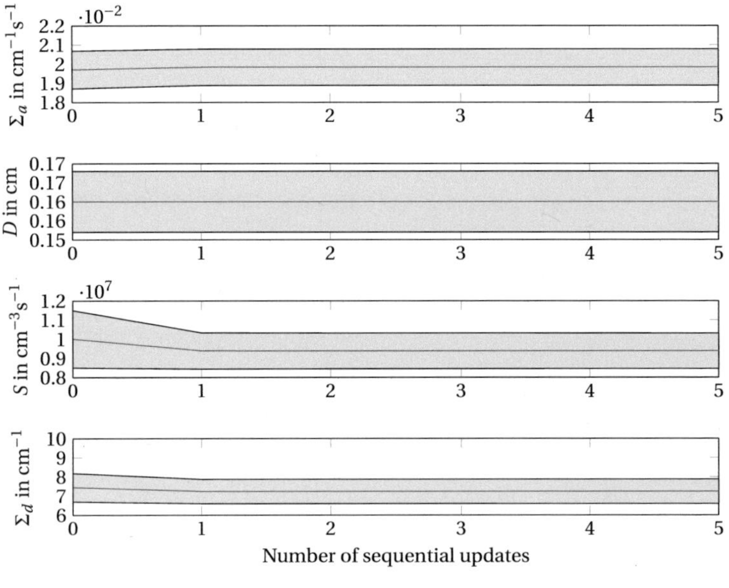

Figure 5.4: Mean and 2σ interval of the calibrated parameters by using the Bayesian update method

The main computational effort in each step depends of the computation of model response $\mathbf{Y}^M$. The algorithm needs to calculate the model at quadrature nodes to compute the coefficients of $\mathbf{Y}^M$. In order to achieve a sufficiently accurate result, the sparse grid GHQ with level of 4 is applied. 681 model evaluations are required for each steps (see example in section 3.4). This leads to 3405 model evaluations for 5 steps. However, it can be observed in figure 5.4,that there are hardly changes of the parameter values.

The update procedure is actually accomplished in 2 steps, which means that only 1362 model evaluations are necessary.

The result of this method is the coefficients of PCE representation of the assimilated parameters $\mathbf{Q}^a$. Using the visualization tool of the developed software, the posterior distribution calculated by the Bayesian update after 5 steps is presented by the histograms and the scatter plots in figure 5.5.

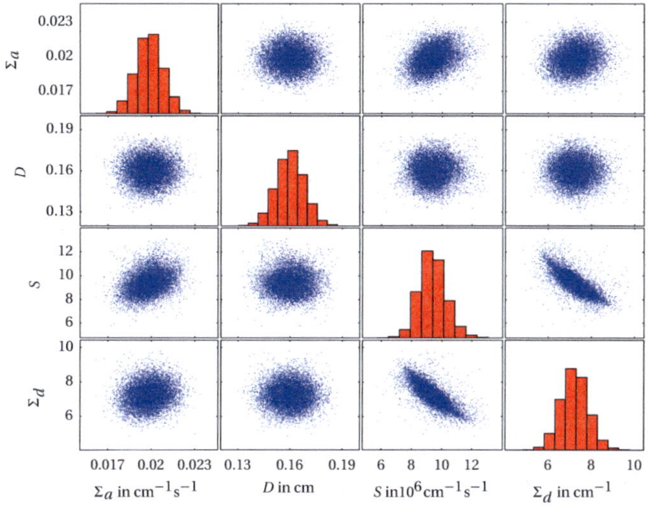

Figure 5.5: Histogram and scatter plots of the samples from the posterior distribution assessed by the Bayesian update method

The means, the standard deviations and the correlation matrix of the calibrated parameters are computed directly from the PCE coefficients by using equation (3.20). All statistical values are summarized in the tables 5.1, 5.2 and 5.3.

gPC surrogate model

As mentioned previously, that the gPC approximation can be used as a surrogate model. The usages of gPC surrogate in uncertainty propagation and sensitivity analysis are demonstrated in previous examples in section 3.4 and 4.2.1 respectively.

Figure 5.6a shows a comparison between the analytical solution and the gPC approximation, generated in example 4.2.1, of order $P = 4$ and $P = 7$ for the parameter Σ_a. Figure 5.6a exhibits that the Legendre gPC are able to approximate the analytical solution well. The approximation is evaluated by the relative error

$$e_{\text{rel}} = \frac{|y^{\text{M}} - y^{\text{gPC}}|}{|y^{\text{M}}|}. \tag{5.47}$$

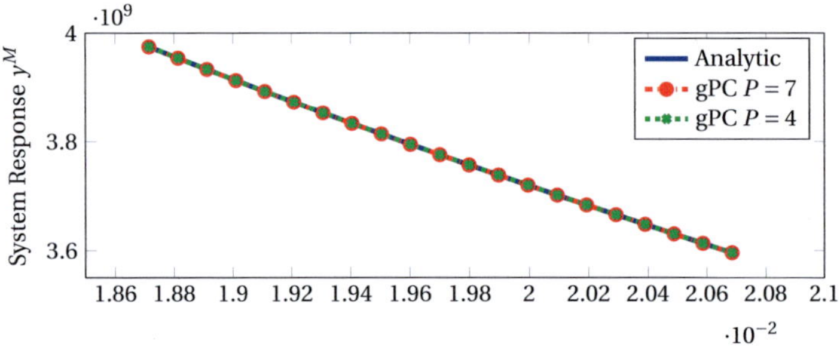

(a) System response by gPC approximation $P = 4$, $P = 7$, and analytical solution

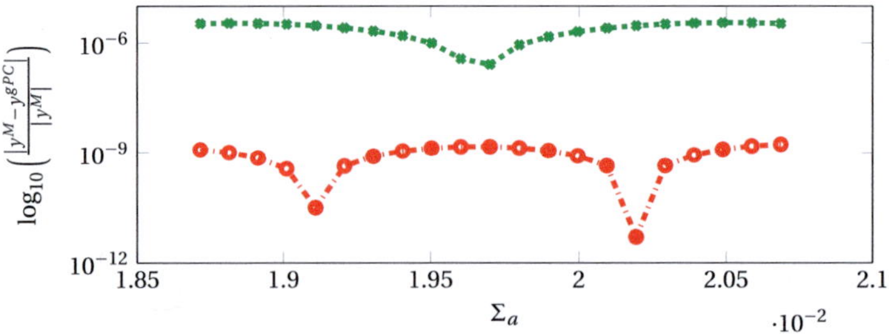

(b) Relative error of the gPC approximation in logarithmic scale

Figure 5.6: Comparison of the gPC approximation $P = 4$ and $P = 7$ with the analytical solution

The relative errors of both gPC approximation are presented in a log scale in figure 5.6b. The relative error of gPC approximation of order $P = 7$ is about 10^{-9}, i.e. the gPC approximation has an extremely small approximation error. The relative error of gPC order $P = 4$ is about 10^{-6}, which much larger than the error of gPC $P = 7$, but still relatively small. The relative error of about 10^{-6} should be accurate enough for this example. Therefore, the gPC order $P = 4$ is applied as surrogate in this example.

The Legendre gPC $\hat{\boldsymbol{y}}^M(\boldsymbol{q}) = \sum_{k=0}^{N_P} \beta_k \Psi_k(\boldsymbol{q})$ generated in example 4.2.1 replaces the full model $\boldsymbol{y}^M$ in the likelihood function (Eq.(5.8)). The posterior is assessed by the MCMC method as before. The posterior distribution result of the batch method with surrogate model is quite similar to the result with the full model. The statistical values calculated from the samples are summarized in the tables 5.1, 5.2 and 5.3 respectively.

Cacuci's BEST-EST method requires a Jacobi matrix $\mathbf{J}$ to calibrate the parameters as shown in Eq.(5.41). Sometimes the calculation of the Jacobi matrix could be difficult, if an analytical formula is not available. As shown in section 4.2, that gPC is able to approximate the Jacobi matrix. The gPC can be utilized as a surrogate model to approximate the system responses and the Jacobi matrix. In this example, the parameters are calibrated with the BEST-EST method by using a gPC approximation constructed in example 4.2.1 as a surrogate model to approximate the system response and the Jacobi matrix $\mathbf{J}$. The statistical values resulted by this BEST-EST method with gPC surrogate are summarized in the tables 5.1, 5.2 and 5.3 respectively.

Discussion

The statistical values, namely the mean values, the standard deviations and the correlation matrix, before and after calibration of all methods are summarized in the tables 5.1, 5.2 and 5.3 respectively. Most results of all calibration methods are in good agreement to each other. The visible differences can be found only in the correlation matrices, however, they have only values of the order below 10^{-2}, which is neglectable. The important correlation pair components, namely (Σ_d,S), (Σ_d,Σ_a) and (Σ_a,S), of all methods deviate very small from each others.

It should be noted that the results from the batch method are calculated from samples, which vary in each run due to the relatively small number of samples (10^4). On the contrary, the Bayesian update with PCE method is deterministic. In the sense that the result does not vary in each run, if the settings such as polynomial order or quadrature level are not changed.

As mentioned previously, the Cacuci's BEST-EST method can be also considered as a Bayesian update method. The BEST-EST approximates the system response with the first-order Taylor series, while the PCE Bayesian update employ the stochastic spectral method. It should be noted that the first-order Taylor approximation has limited usage if the model is highly nonlinear. The PCE update can work with the highly nonlinear system by using a higher order polynomials as demonstrated in [Paj12].

Calibration Method	Expectation of			
	Σ_a^{cal} / $cm^{-1}s^{-1}$	D^{cal} / cm	$S^{cal}/10^6$ $cm^{-3}s^{-1}$	Σ_d^{cal} / cm^{-1}
BEST-EST [Cac14a]	0.0198	0.160	9.395	7.238
Batch (full model)	0.0198	0.160	9.383	7.260
PCE Bayesian update (full model)	0.0198	0.160	9.383	7.234
BEST-EST with gPC surrogate	0.0198	0.160	9.395	7.238
Batch with surrogate model	0.0198	0.160	9.420	7.242
Before Calibration	0.0197	0.160	10.000	7.438

Table 5.1: Expectation values of the calibrated parameters by different approaches

Calibration Method	Standard deviation of			
	$\Sigma_a^{cal}/10^{-4}$ $cm^{-1}s^{-1}$	$D^{cal}/10^{-2}$ cm	$S^{cal}/10^5$ $cm^{-3}s^{-1}$	$\Sigma_d^{cal}/10^{-1}$ cm^{-1}
BEST-EST [Cac14a]	9.512	0.800	9.396	6.355
Batch (full model)	9.685	0.815	9.528	6.673
PCE Bayesian update (full model)	9.515	0.800	9.288	6.388
BEST-EST with gPC surrogate	9.512	0.800	9.396	6.355
Batch with surrogate model	9.360	0.790	9.202	6.412
Before Calibration	9.850	0.800	15.000	7.438

Table 5.2: Standard deviations of the calibrated parameters by different approaches

No significant difference between the result by the full model and by the gPC surrogate model can be observed from the tables. The BEST-EST with gPC gives the exact result of the expectations and the standard deviations as the BEST-EST with full model. The results of the batch methods with gPC deviate a little from the full model. However, MCMC is a sampling-based method, which results vary due to the low number of samples. It shows that the gPC approximation can be used efficiently as a surrogate in the Bayesian inverse problem solution.

Figure 5.2 and the values in all table reveal that the changes of the parameters values is in the following order: S, Σ_d, Σ_a and D, whereas the parameter D is nearly unchanged. This result is also in good agreement with the results of the sensitivity analysis in the example in section 4.4.1 as the global sensitivity analysis provide the same ranking.

Lastly, the model is calculated with the posteriors of the calibrated parameters. The results are illustrated in figure 5.7. The posterior from all calibration methods deliver similar results of the expectations and the confidence intervals of the system response. As there are some small variations in the distributions, some small differences appear between the predictions of each method. However, they are in good agreement with the measurements, which affirm the resulted posteriors of the calibrated parameters by all mentioned methods.

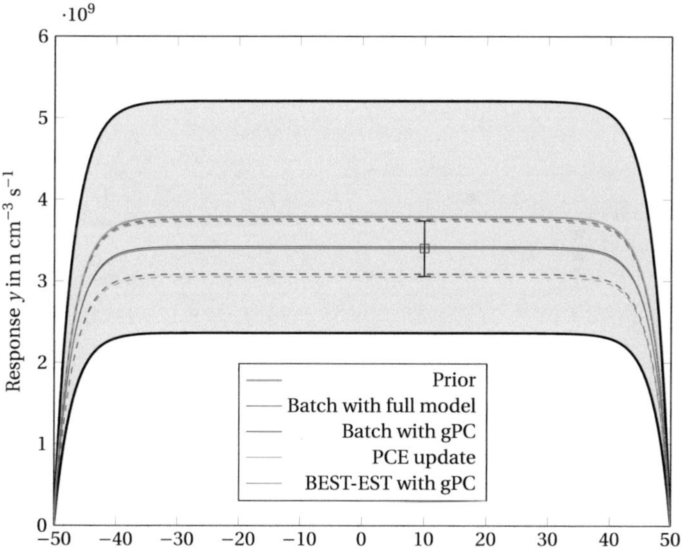

Figure 5.7: System response of the neutron diffusion model with the posterior distribution of the calibrated parameters. (solid line = expectation, dashed line = 2σ interval, □ denotes the measurement)

Correlation Matrix $\mathbf{R}_{\boldsymbol{Q}}^{\mathrm{cal}}$ for each Calibration Method
BEST-EST [Cac14a]
$\mathbf{R}_{\boldsymbol{Q}}^{\mathrm{cal}} = \begin{bmatrix} 1.000 & -3.942\times10^{-7} & 3.349\times10^{-1} & 1.637\times10^{-1} \\ -3.942\times10^{-7} & 1.000 & 1.823\times10^{-6} & 8.912\times10^{-7} \\ 3.349\times10^{-1} & 1.823\times10^{-6} & 1.000 & -7.570\times10^{-1} \\ 1.637\times10^{-1} & 8.912\times10^{-7} & -7.570\times10^{-1} & 1.000 \end{bmatrix}$
Batch with full model
$\mathbf{R}_{\boldsymbol{Q}}^{\mathrm{cal}} = \begin{bmatrix} 1.000 & -3.975\times10^{-7} & 3.153\times10^{-1} & 1.437\times10^{-1} \\ -3.975\times10^{-7} & 1.000 & -1.198\times10^{-2} & 2.638\times10^{-2} \\ 3.153\times10^{-1} & -1.198\times10^{-2} & 1.000 & -7.646\times10^{-1} \\ 1.437\times10^{-1} & 2.638\times10^{-2} & -7.646\times10^{-1} & 1.000 \end{bmatrix}$
PCE Bayesian update with full model
$\mathbf{R}_{\boldsymbol{Q}}^{\mathrm{cal}} = \begin{bmatrix} 1.000 & -2.235\times10^{-5} & 3.188\times10^{-1} & 1.561\times10^{-1} \\ -2.235\times10^{-5} & 1.000 & -1.070\times10^{-4} & 9.769\times10^{-6} \\ 3.188\times10^{-1} & -1.070\times10^{-4} & 1.000 & -7.487\times10^{-1} \\ 1.561\times10^{-1} & 9.769\times10^{-6} & -7.487\times10^{-1} & 1.000 \end{bmatrix}$
BEST-EST with gPC surrogate model
$\mathbf{R}_{\boldsymbol{Q}}^{\mathrm{cal}} = \begin{bmatrix} 1.000 & -3.876\times10^{-7} & 3.335\times10^{-1} & 1.637\times10^{-1} \\ -3.876\times10^{-7} & 1.000 & 1.793\times10^{-6} & 8.762\times10^{-7} \\ 3.335\times10^{-1} & 1.793\times10^{-6} & 1.000 & -7.570\times10^{-1} \\ 1.637\times10^{-1} & 8.762\times10^{-7} & -7.570\times10^{-1} & 1.000 \end{bmatrix}$
Batch with gPC surrogate model
$\mathbf{R}_{\boldsymbol{Q}}^{\mathrm{cal}} = \begin{bmatrix} 1.000 & 1.728\times10^{-5} & 3.064\times10^{-1} & 1.368\times10^{-1} \\ 1.728\times10^{-5} & 1.000 & -4.280\times10^{-2} & 5.119\times10^{-2} \\ 3.064\times10^{-1} & -4.280\times10^{-2} & 1.000 & -7.582\times10^{-1} \\ 1.368\times10^{-1} & 5.119\times10^{-2} & -7.582\times10^{-1} & 1.000 \end{bmatrix}$
Before calibration (uncorrelated)
$\mathbf{R}_{\boldsymbol{Q}} = \begin{bmatrix} 1 & 0 & 0 & 0 \\ 0 & 1 & 0 & 0 \\ 0 & 0 & 1 & 0 \\ 0 & 0 & 0 & 1 \end{bmatrix}$

Table 5.3: Correlation matrices of the calibrated parameters by different approaches

6

Application to a Glass Forming Process

In a wide variety of industrial processes the underlying physical phenomena have to be regarded as spatially distributed. Manufacturing of glass is an example of such a process. Forming of glass is a very complex rheological forming process. In the manufacture of some glass products such as optical fibers and its pre-products, the precise control of the production is critical to the final quality of the product. It is of significant interest to understand the effects that cause changes in the quality of the products. The modeling and simulation can play an important role to deal with this issues.

However, the applications of computational models to investigate the various aspects of glass manufacturing processes are mostly based on the assumption that the parameters governing the transport processes are deterministic. Considerable uncertainty is inherent in the process, arising from many sources such as operating parameter fluctuations, inaccuracies in process control, empirical determination of the transport parameters, and environmental uncertainties. Due to all these uncertainties, the computational model may provide solutions, which do not agree very well with the measurements of the real process.

The capabilities of the proposed framework for analysis and identification using Bayesian UQ based on gPC has been demonstrated with academic numerical examples in the previous chapters. As described in chapter 2, the framework is composed mainly of three procedures, namely uncertainty propagation, sensitivity analysis and Bayesian inference. The first two are used to support in understanding of the behavior of systems, while the latter is used to solve the discrepancy between the computational model and the measurements of real processes.

In the following, the application of the proposed framework to a model of a real glass forming process by means of a study case is presented. This model comes from real-world industry production, which was developed under a R&D cooperation between Fraunhofer IOSB and an industrial partner. It should be noted that some information about the application described in this chapter has to be suppressed because of a confidentiality agreement between Fraunhofer IOSB and the industrial partner.

6.1 Glass-forming process

In industrial-scale glass manufacturing, glass forming operations vary among the various sectors such as fibers, tubes, glass containers and TV panels [Cho10]. The glass forming process considered in this thesis is glass production by drawing a continuous stream of glass. It is the production of cylindrical glass tubes and rods, so-called *preforms*, from thick glass cylinders. A preform is the source material which will be drawn to form the long, thin optical fiber.

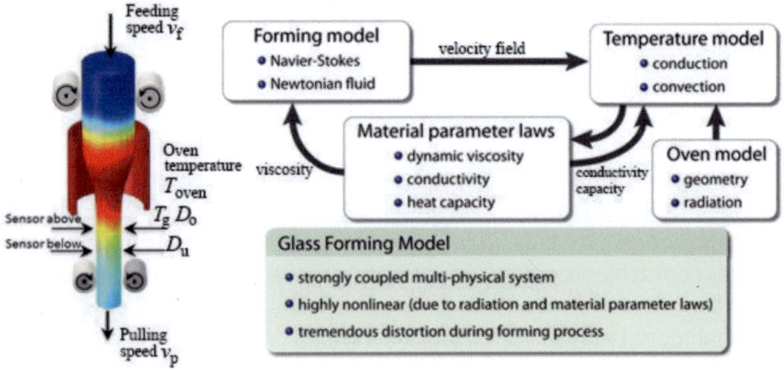

Figure 6.1: Schematic illustration of the industrial glass forming process (left) and structure of the glass forming model with its nonlinearities (right)

The forming process of a preform is a complex rheological process. The process involves a wide temperature range and is characterized by large deformations. The process setup is visualized in figure 6.1 (left). The thick glass cylinder is fed with slow velocity v_{feed}, called feeding speed, in a furnace where it is heated up. As the glass reaches the forming temperature, it starts softening and acting like a liquid. Below the furnace the glass is pulled with a pulling velocity v_{pull}, which is usually higher than the feeding speed, and it results in a thin glass rod (respectively tube) as a consequence. The process parameters are regulated by means of a control algorithm, to which measurements from sensors installed in the process are fed.

During drawing, temperature and velocities within the glass body change along radius, angel and vertical axis. The quality of the preform strongly depends on the temperature and velocity distribution throughout the cylinder. The temperature and velocity distributions depend on the process parameters on the other hand. Because the properties of the optical fiber rely on the quality of the preform, the process parameters need to be regulated accurately in order to achieve the required attributes of the preform. For calculating optimal dynamical parameter settings, a mathematical model of the process is needed.

The mathematical model formulation of different glass forming processes can be found in literature, for example [Loc02], [Far11], and [Cho10]. One of the most discussed glass forming process is the modeling of the optical fiber drawing process, which relates to the preform glass forming process model in this work from physics viewpoint. Based on fundamental physical laws, the flow and heat transfer phenomena in the glass forming process, both optical fiber drawing and preform manufacturing, can described with the conservation laws of mass, momentum and energy. The distinctions of the geometry dimension and the process setup for preform manufacturing make the glass forming model difference to the optical fiber drawing process. Because in general optical fibers are drawn with a very small diameter, some simplifications can be assumed in the modeling, which cannot be assumed in case of the preform forming process model. The description of the preform process model considered in this thesis is discussed in the following subsection.

6.1.1 Mathematical model of glass forming process

The main physical phenomena of the glass forming process arise from the radiation, heat convection, and fluid dynamics. Basically, the model can be divided in two main parts, i.e. (a) the glass flow and (b) the heat transfer in the glass and from the furnace to the glass (see figure 6.1 (right)). The glass is assumed to be mechanically incompressible and behaves as a Newtonian fluid, which hold for the considered process. The effect of thermal expansion is assumed to be insignificant and is neglected in this consideration. With these assumptions, the conservation laws of mass, momentum and energy including the constitutive relation formulate the system of PDEs of the process in Eulerian formalism:

$$\frac{\partial \rho}{\partial t} + \nabla^T \left(\rho \boldsymbol{v}\right) = 0, \tag{6.1}$$

$$\frac{\partial \rho \boldsymbol{v}}{\partial t} + \left(\boldsymbol{v}^T \nabla\right) \rho \boldsymbol{v} = \nabla^T \boldsymbol{\sigma}(\boldsymbol{v},p) + \rho g, \tag{6.2}$$

$$\rho c_p(T) \frac{\partial T}{\partial t} + \left(\boldsymbol{v}^T \nabla\right) \left(\rho c_p(T)\, T\right) = \nabla^T \left(\lambda(T) \nabla T\right). \tag{6.3}$$

In the equation it denotes $\boldsymbol{v}$: velocity vector, T: temperature of the glass, c_p: specific heat capacity, λ: effective heat transfer coefficient (considering radiative heat transfer in a simplified way), ρ: density of the glass, g: gravitational acceleration and $\boldsymbol{\sigma}$: Cauchy stress tensor, which is defined for Newtonian incompressible fluid as

$$\boldsymbol{\sigma}(\boldsymbol{v},p) = -p\mathbf{I} + \eta(T)\left(\nabla \otimes \boldsymbol{v} + (\nabla \otimes \boldsymbol{v})^{\mathrm{T}}\right), \tag{6.4}$$

where η is the dynamic viscosity, p is the pressure and $\otimes$ denotes the tensor product.

All variables are described in the time domain $\mathcal{T} = [t_{\text{start}}, t_{\text{end}}]$ and in the spatial space $\mathcal{G}$. The PDEs are non-linear regarding to the convection term $\left(\boldsymbol{v}^T \nabla\right) \rho \boldsymbol{v}$ and the material parameters λ, c_p, and η which depend on the temperature T. The material properties of the glass are described in section 6.1.3. Because of the cylindrical form of the preform and the tube, the cylindrical coordinate system $\boldsymbol{r} = (r, \varphi, z)^T$ is used to described the position in the spatial space. The spatial space is defined for

- $z \in [z_{\text{end}}, z_{\text{start}}]$,
- $\varphi \in [0, 2\pi]$ and
- $r \in [R_{\text{in}}(z), R_{\text{out}}(z)]$ ($R_{\text{in}}(z) = 0$ in case of a preform).

As the glass cylinder is deformed during the process, the spatial domain $\mathcal{G}$ of the model is changed in time. The geometry of the glass forming model before the deformation ($t_{\text{start}} = 0$ s) is shown in figure 6.2 and an example of the geometry after the deformation is illustrated qualitatively in figure 6.3.

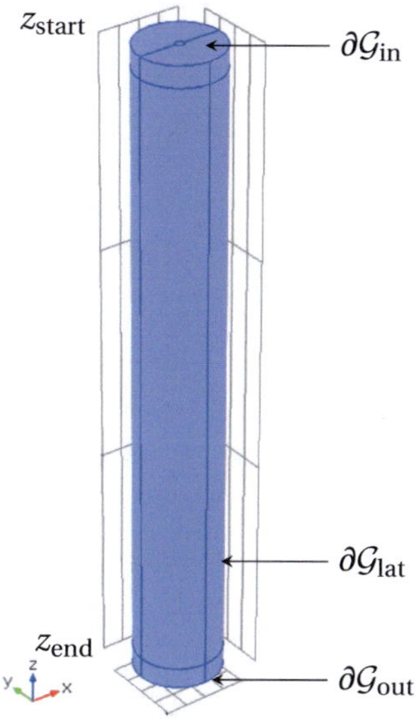

Figure 6.2: Geometry of the glass model before the deformation and the boundary conditions

6.1.2 Boundary condition

In order to solve the PDEs, it is necessary to formulate the interaction phenomena between the flowing system and the surrounding into boundary conditions. The boundaries of the the glass forming model can be categorized into three domains, namely the lateral boundary $\partial\mathcal{G}_{\text{lat}}$, the inlet boundary $\partial\mathcal{G}_{\text{in}}$ and the outlet boundary $\partial\mathcal{G}_{\text{out}}$, as shown in figure 6.2. One of the key interactions is the energy flow at the lateral boundary, which is described by the *oven model*.

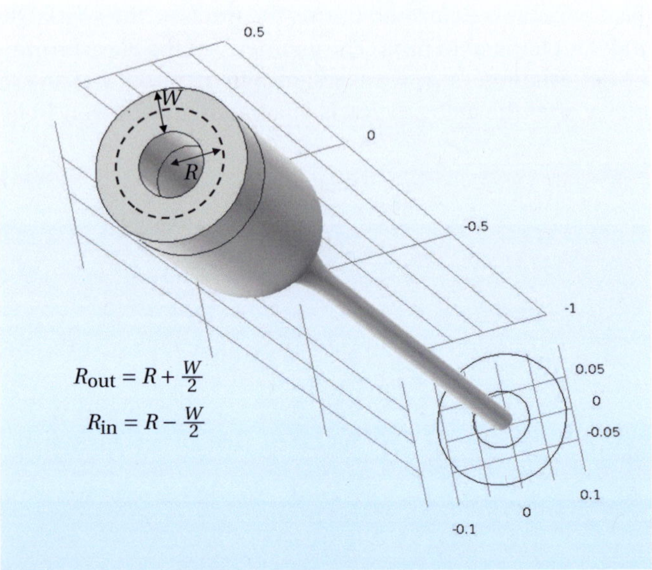

Figure 6.3: Geometry of the glass tube after the deformation (spatial coordinate in m)

Oven model

The oven model is characterized by the temperature distribution in the oven and the thermal radiative heat exchange between the glass and the oven. With the Stefan-Boltzmann laws, the relation of the radiative heat flux q''_{rad} at boundary is described as:

$$q''_{\text{rad}} = \epsilon\sigma_B \left(T^4_{\text{oven}}(z) - T^4\right) = -\boldsymbol{n}^T (\lambda \nabla T), \tag{6.5}$$

where ϵ: is the emissivity of the glass, σ_B: Stefan-Boltzmann constant, $T_{\text{oven}}(z)$: oven temperature distribution in the oven. In this work, the oven temperature is assumed to be rotation-symmetric and therefore is only a function of the z-coordinate. The oven temperature $T_{\text{oven}}(z)$ is illustrated qualitatively in figure 6.4.

$T_{\text{oven}}(z)$ is assumed to be Gauss-like curve depending on some specific process parameters such as ambient temperature T_{amb} or maximal oven temperature $T_{\text{oven\_max}}$. The curve of $T_{\text{oven}}(z)$ is described by the function

$$T_{\text{oven}}(z) = \begin{cases} T_{\text{amb}} + \left(T_{\text{oven\_max}} - T_{\text{amb}}\right) \exp\left(\frac{-|z-z_0|^{a_{\text{above}}}}{b_{\text{above}}}\right), & \text{for } z \geq z_0 \\ T_{\text{amb}} + \left(T_{\text{oven\_max}} - T_{\text{amb}}\right) \exp\left(\frac{-|z-z_0|^{a_{\text{below}}}}{b_{\text{below}}}\right), & \text{for } z < z_0 \end{cases} \tag{6.6}$$

The parameters a_{above}, b_{above}, a_{below} and b_{below} are fitted by specifying the mentioned temperatures and additionally the temperatures at some arbitrary points z, e.g. at the end of the furnace.

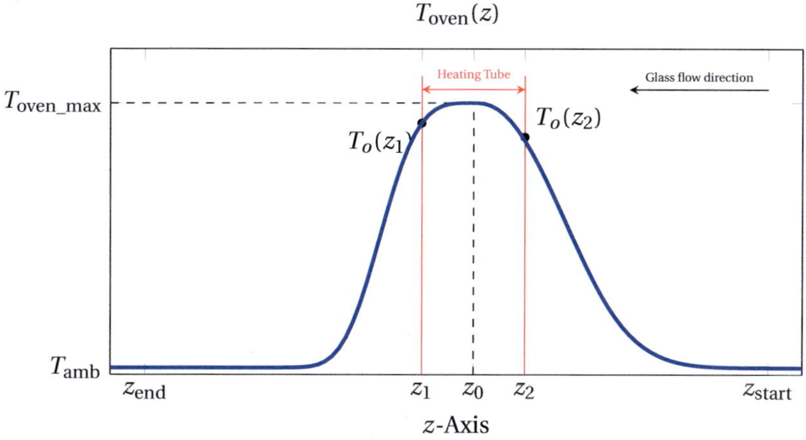

Figure 6.4: Dependence of oven temperature respect to the z-coordinate

The glass cylinder with the temperature T_{cyl} is fed at the upper boundary $z = z_{\text{start}}$ with the velocity v_{f}. Therefore, at the upper boundary the Dirichlet boundary condition is defined by fixing temperature of glass cylinder T_{cyl}. At the outlet boundary, the convection-dominated boundary condition with the ambient temperature T_{amb} is applied. This condition states that the heat transfer occurring across the boundary is only by convection. The temperature gradient in the normal direction, namely the conductive heat flux $\boldsymbol{q}''_{\text{cond}}$, is zero.

$$T = T_{\text{cyl}} \qquad \text{for } \boldsymbol{r} \in \partial\mathcal{G}_{\text{in}} \tag{6.7}$$

$$\boldsymbol{q}''_{\text{cond}} = 0 \qquad \text{for } \boldsymbol{r} \in \partial\mathcal{G}_{\text{out}} \tag{6.8}$$

For the flow, the Navier-Stokes with free surface flow allows the calculation of the geometry deformation. The formulation of the boundary condition for the free surface flow can be found in [Gan06]. The geometry of the outer lateral boundary moves with the interface velocity $\boldsymbol{v}_{\text{interface}}$. At the lateral boundary the kinetic boundary condition for the free surface flow is applied as followed.

$$\boldsymbol{n}^T \boldsymbol{\sigma}(\boldsymbol{v},p)\,\boldsymbol{n} = \sigma_s \mathcal{K} \qquad \text{for } \boldsymbol{r} \in \partial\mathcal{G}_{\text{lat}} \tag{6.9}$$

$$\boldsymbol{\tau}_i^T \boldsymbol{\sigma}(\boldsymbol{v},p)\,\boldsymbol{n} = 0 \qquad \text{for } \boldsymbol{r} \in \partial\mathcal{G}_{\text{lat}} \tag{6.10}$$

σ_s the surface tension coefficient and $\mathcal{K}$ is the sum of the principal curvatures. Further $\boldsymbol{\tau}_i, i = 1,2$ are tangential vectors on the free surface. The shear stress (eq.(6.10)) vanishes at a free liquid surface. The normal stress boundary condition at a fluid-fluid interface actually determines the curvature of the interface at the point in question. However, due to the geometry scale, the pressure caused by the surface tension can be neglected compared to atmosphere pressure. Apart from the kinetic boundary condition, the kinematic condition at the free surface

$$\boldsymbol{v}^T \boldsymbol{n} = \boldsymbol{v}_{\text{interface}}^T \boldsymbol{n} \qquad \text{for } \boldsymbol{r} \in \partial\mathcal{G}_{\text{lat}} \tag{6.11}$$

has to be satisfied, i.e., the normal velocity of the fluid at the interface should be equal to the normal velocity of the interface. Besides the free surface flow condition at the lateral boundary, the Dirichlet boundary condition is applied at the inlet and outlet boundaries. The velocities at inlet and outlet boundaries are set to equal the feeding speed and the pulling speed as:

$$\boldsymbol{v}(\boldsymbol{r} \in \partial\mathcal{G}_{\text{in}}) = \boldsymbol{v}_{\text{feed}} = -v_{\text{f}}\boldsymbol{e}_z, \tag{6.12}$$

$$\boldsymbol{v}(\boldsymbol{r} \in \partial\mathcal{G}_{\text{out}}) = \boldsymbol{v}_{\text{pull}} = -v_{\text{p}}\boldsymbol{e}_z. \tag{6.13}$$

6.1.3 Fluid physical properties of glass

Aside from the process parameters, the forming process depends on the material properties, as described in eq.(6.1)-eq.(6.4) as well. The material parameters of glass vary strongly with the relevant temperature range during the forming process. The nonlinear materiel parameter laws of glass are described in the following:

Heat Capacity The heat transfer in the glass as stated in equation (6.3) principally relies on the heat conductivity and heat capacity parameters. The specific heat capacity c_p is in general slightly dependent on the temperature. According to measurements of our industry partner, the following relation is employed in this paper.

$$c_p(T) = c_{p_1} + c_{p_2}\left(T - c_{p_3}\right) \tag{6.14}$$

Although two parameters would be sufficient to represent the linear dependence, an approach with three parameters in eq. (6.14) is chosen because it is easier to be interpreted for the modeler. The relation between the specific heat capacity and the temperature is shown in figure 6.5a.

Heat Conductivity In modeling of heat transfer in the glass forming process, the heat flux is the result of the contribution of the thermal conduction and the radiation. The calculation of the conductivity due to the radiation is often complicated. The Rosseland approximation [Ros24], where the radiative conductivity λ_{rad} is assumed to be proportional to T^3, is commonly applied for the radiative conductivity in glass models [Gro11], [Pye05], [Cho05]. According to our industry partner's measurements and experiences, the modified relation of the effective heat conductivity is used as

$$\lambda(T) = \underbrace{\lambda_1}_{\text{thermal conductivity}} + \underbrace{\lambda_2 \cdot T^{\lambda_3}}_{\text{radiative conductivity}} . \tag{6.15}$$

This relation has the same structure as the Rosseland approximation in case of $\lambda_3 = 3$. The dependence between the effective heat conductivity and the temperature is presented in figure 6.5b.

Dynamic Viscosity The glass is a rheological material exhibiting viscoelastic properties. However, the motion of glass is dominated by viscous flow and the influence of elastic effects is often neglected [Gro11], which is also assumed in this thesis. The viscous properties of glass can be determined by the dynamic viscosity. In case of soda lime silica glasses, the dynamic viscosity is dependent on the temperature and the shear rate [Sim89]. As the viscosity range considered in this application depends mainly on the temperature, the dynamic viscosity is formulated as a function of temperature by means of the available experimental measurements. The range of the viscosity of glass for varying temperature is relatively large. Typically the temperature dependence of the viscosity of glass about the melting temperature is given by the Vogel-Fulcher-Tammann(VFT) relation [She05]. As it can be observed in figure 6.5c, the viscosity value by VFT relation provide an extremely large value of viscosity at low temperature, which could be difficult for the numerical solver of the PDEs. In order to facilitate the numerical computation, the maximal value of dynamic viscosity is limited by using the *modified relation*

$$\log_{10}\left(\eta(T)\right) = \eta_1 + \eta_2 \cdot \tanh\left(\eta_3(T - \eta_4)\right). \tag{6.16}$$

A typical curve of the dependence between the viscosity and the temperature using in this application is illustrated in figure 6.5c.

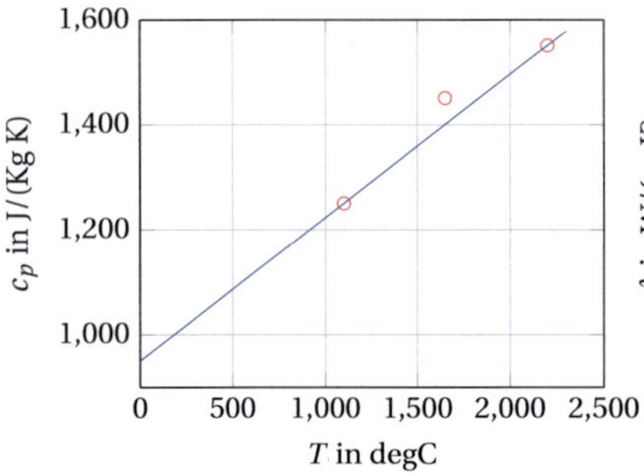

(a) Specific heat capacity

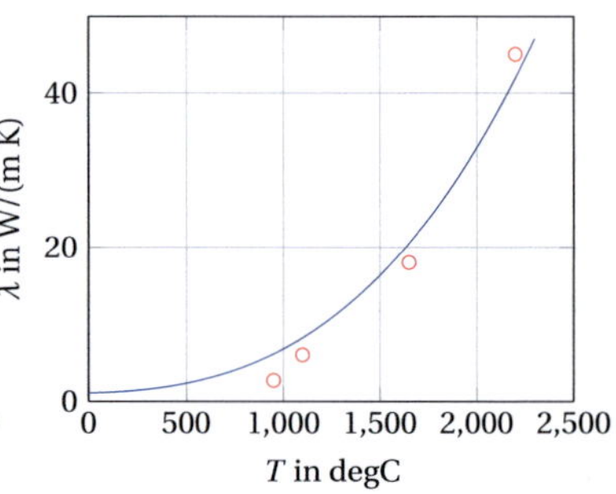

(b) Heat conductivity

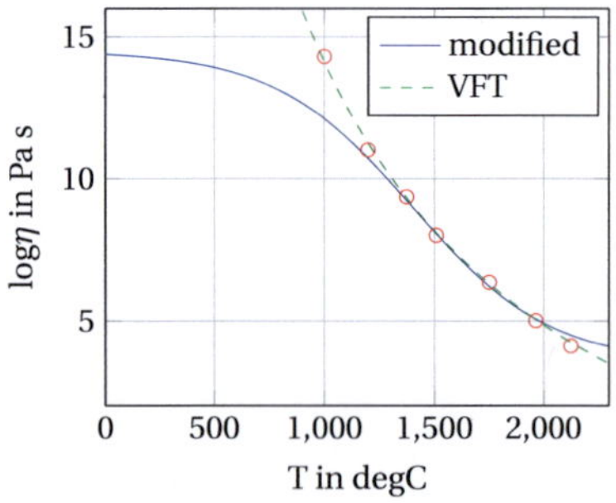

(c) Viscosity

Figure 6.5: Temperature dependency of material parameters, measurements (red circles) and their interpolated function (blue line)

All parameters λ_1, λ_2, λ_3, c_{p_1}, c_{p_2}, c_{p_3} as well as η_1 till η_4 for the material model are usually determined by the modeler, based on available information, some empirical knowledge or experimental measurement data, and therefore they are uncertain. The uncertainty of the parameters could provide the discrepancy between the model predictions and the measurement data as discussed later in section 6.2. In this chapter, a parameter calibration method based on measurement data and model predictions by means of Bayesian inverse problem formulation is proposed.

6.1.4 Implementation

In general, there is no analytical solution for the initial-boundary value problem (IBVP), especially for such a complex nonlinear system as the glass forming process model. Therefore, the IBVP should be solved approximately by numerical method such as Finite Element Method (FEM) or Finite Volume Method (FVM). The FEM approximates the solution of the IBVP by discretizing the spatial space as discussed in section 2.2.1.

The PDEs and the boundary conditions discussed previously are defined in general in the three dimensional (3D) or two dimensional (2D) spatial space $\mathcal{G}$. The 3D model is demanded e.g. in case of analysis of oval deformations of the preform due to hot spots in the furnace. Nevertheless, the 3D model requires tremendous computational effort. Because of the high computational effort of 3D, one tries to avoid the full computation, if it is not necessary.

In case of fundamental analysis, such as analysis of the diameter or temperature, where the variables do not depend on the angle φ, the 2D models can be deduced from the axisymmetric assumption. All variables in the model, hence, depend on the spatial coordinate $\boldsymbol{r} = (r, z)^T$ with the radius r and the height z. Simulation of 2D models requires quite less effort than the full 3D model, therefore the two-dimensional (2D) model is rather preferred for the fundamental analysis in general.

In the course of this thesis, all partial differential equations, the boundary conditions and the constitute equations of the glass forming process model discussed in the last section are implemented in the commercial FEM software COMSOL version 5.2. Figure 6.6 shows one of the results of the 2D model. More details about the glass forming Finite Element model can be found in the previous works [Ber06], [Saw09], [Ja11], [Ja13b] and [Ja15b] at the IOSB.

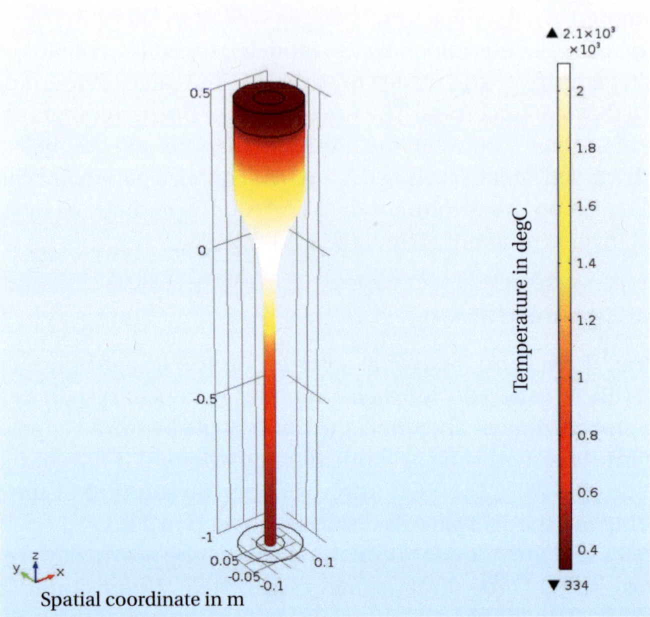

Figure 6.6: Temperature distribution of the forming process of a glass rod by 2D-FEM model with 3D visualization

Implementation of moving boundaries

The main challenge of the free surface flow computation in 3D and 2D models is the moving boundary and the deformed mesh. In continuum mechanics there are two descriptions of spatial space, namely the *Lagrangian* and the *Eulerian* descriptions. The Eulerian description represents a field as a function of position $\boldsymbol{x}$, while the Lagrangian description represents a field as a function of material coordinate $\boldsymbol{x}_0$. The Lagrangian description is applied mostly in solid mechanics problems, as the mesh can move with the deformation of the material. In most cases of fluid flow problems, the Eulerian description is preferred. For such descriptions the mesh remains fixed while the material passes through it. With the Eulerian description, it is difficult to track the moving boundary.

The *Arbitrary Lagrangian-Eulerian* (ALE) method was developed in attempt to combine the advantages of the Lagrangian description and Eulerian description. The application of ALE method to the free surface flow have been presented in many publications e.g. [Don82], [Nob01], [Wan97]. In the FEM implementation in this thesis, the tracking of the moving boundary succeed by using *Arbitrary Lagrangian-Eulerian* (ALE) method.

1D model simplification

In case of rapid test, the one-dimensional (1D) model can be used. The 1D model, sometimes called *Trouton* model, [How94, Loc02] is derived from the thin-layer flow assumption. It is supposed, that the radius of the glass cylinder R is very small compared to the typical length L in z-direction ($R/L \ll 1$). In the 1D model, the variables are only the function of space $\boldsymbol{r} = z$ and time t. The glass *tube* version of the *Trouton* model is:

$$\frac{\partial}{\partial t}(RW) + \frac{\partial}{\partial z}(v_z RW) = 0, \tag{6.17}$$

$$\frac{\partial}{\partial z}\left(3\eta(T)RW\frac{\partial v_z}{\partial z}\right) = -\rho g RW, \tag{6.18}$$

$$\frac{\partial R^2}{\partial t} + \frac{\partial}{\partial z}\left(v_z R^2\right) = \frac{p_s}{\eta(T)}\frac{R}{2W}\left(R^2 - \frac{W^2}{4}\right), \tag{6.19}$$

$$A\rho c_p(T)\left(\frac{\partial T}{\partial t} + v_z\frac{\partial T}{\partial z}\right) = \frac{\partial}{\partial z}\left(A\lambda(T)\frac{\partial T}{\partial z}\right) + 2\pi R_{\text{out}} q''_{\text{rad}}. \tag{6.20}$$

where R is the nominal radius of the glass tube (see figure 6.3), W is the wall thickness of the tube, $R_{\text{out}} = R + W/2$ is the outer radius of the tube, v_z is the velocity of the glass in z-direction, p_s is the applied pressure inside the tube, which may vary in the drawing direction and q''_{rad} is the radiative heat flux as eq.(6.5). The cross section area A of the tube can be calculated from:

$$A = \pi R_{\text{out}}^2 - \pi R_{\text{in}}^2, = \pi\left(R + \frac{W}{2}\right)^2 - \left(R - \frac{W}{2}\right)^2, \tag{6.21}$$

$$A = \pi\,(2RW). \tag{6.22}$$

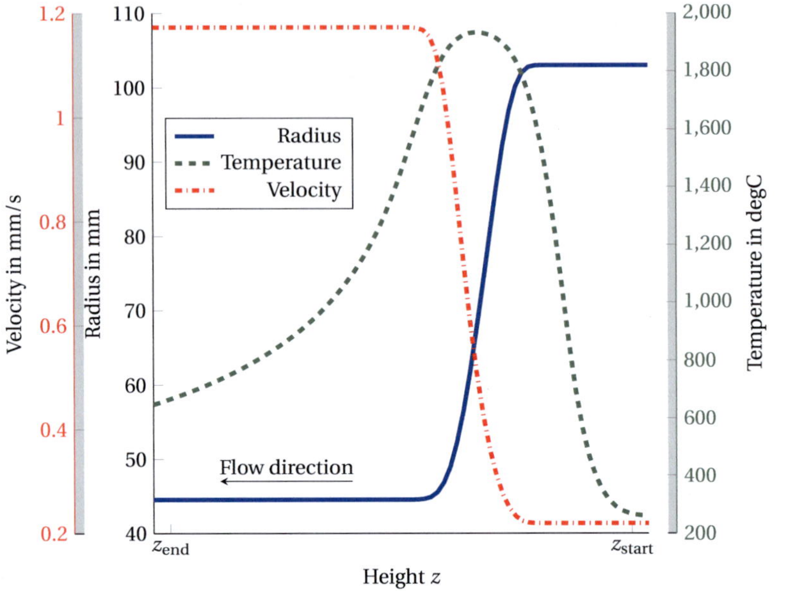

Figure 6.7: Solution of the 1D model simulation of glass rod along the z coordinate. (The values of z are intentionally suppressed due to a confidentiality obligations)

The *preform* or the *glass rod* can be seen as a special case of the *tube* version, where $W = 2R$. The glass *rod* version of the *Trouton* model is then:

$$\frac{\partial A}{\partial t} + \frac{\partial}{\partial z}(v_z A) = 0, \tag{6.23}$$

$$\frac{\partial}{\partial z}\left(3\eta(T) A \frac{\partial v_z}{\partial z}\right) = -\rho g A, \tag{6.24}$$

$$A\rho c_p(T)\left(\frac{\partial T}{\partial t} + v_z \frac{\partial T}{\partial z}\right) = \frac{\partial}{\partial z}\left(A\lambda(T)\frac{\partial T}{\partial z}\right) + 2\pi R_{\text{out}} q''_{\text{rad}}, \tag{6.25}$$

where A is the cross section area of the glass rod. The elaborated derivation of the *Trouton* model can be found in [How94] and [Loc02]. In order to solve these systems of partial differential equations, the boundary condition at $z = z_{\text{end}}$ and $z = z_{\text{start}}$ is required.

In respect of the production process setting, the boundary conditions of the 1D model are assumed as follows:

$$R(z_{\text{start}}) = R_{\text{cyl}}, \tag{6.26}$$

$$W(z_{\text{start}}) = W_{\text{cyl}}, \tag{6.27}$$

$$v_z(z_{\text{start}}) = v_{\text{feed}}, \tag{6.28}$$

$$v_z(z_{\text{end}}) = v_{\text{pull}}, \tag{6.29}$$

$$T(z_{\text{start}}) = T_{\text{cyl}}, \tag{6.30}$$

$$q_{\text{cond}}(z_{\text{end}}) = 0. \tag{6.31}$$

where R_{cyl} and W_{cyl} denote the nominal radius and the wall thickness at the inlet. An example of a result of the 1D model is presented in figure 6.7.

Difference between 1D and 2D

The *Trouton* model is a rough approximation of the glass forming process. It is typically applied to an optical fiber production process, where the radius R is very small compared to the length L. This is, however, not the case concerning the model considered in this thesis. The radius of the preform is small but not small enough to provide a good approximation in the forming zone of the glass.

In the 1D model, it is assumed that the variables do not vary in radial direction. In reality the temperature in the forming zone is not only varying with height z, but also with radius r as demonstrated in figure 6.8. As a consequence, the flow velocity is also varying with r and z as presented by the radial component of the velocity in figure 6.9. Due to this structural inaccuracy of the model the resulting forming behavior calculated by the 1D Trouton model differs from the 2D and 3D model.

The difference of the glass radius between the 1D and 2D model solution is demonstrated in figure 6.10. In the real process, the shrinkage ΔD, which is the difference of diameters between the two sensor positions z_{s1} and z_{s2}, is an important measurement characterizing the status of the production process. It can be observed in figure 6.10, that the 1D model underestimates the shrinkage of ca. 2 mm compared to the 2D model. The shrinkage differences between the 1D and 2D models of the nominal value $\Delta D =$ 4-5 mm is a considerably large error.

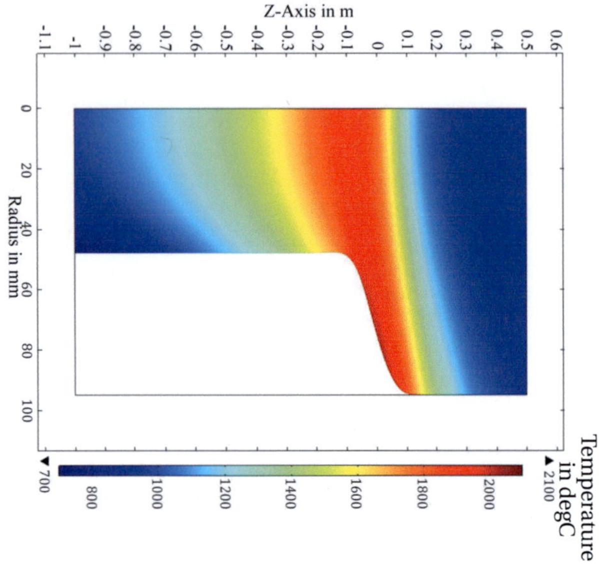

Figure 6.8: Temperature distribution in the 2D model

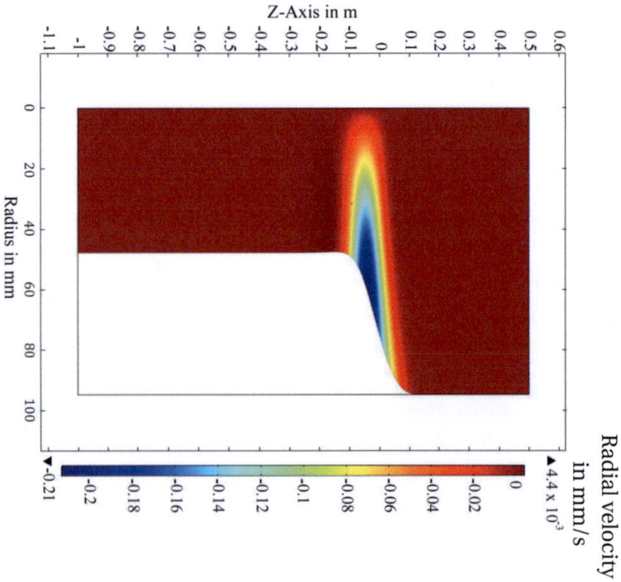

Figure 6.9: Radial velocity distribution in the 2D model

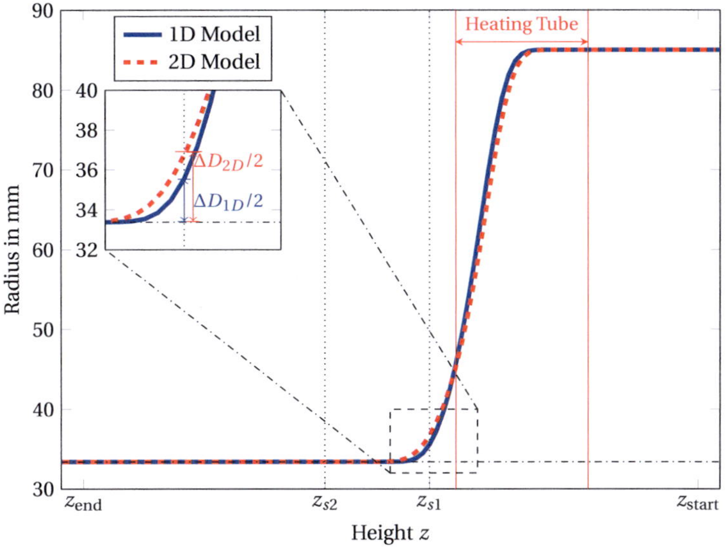

Figure 6.10: Discrepancy between the 1D and 2D model (The values of z are intentionally suppressed due to a confidentiality obligations)

This error affects equation (6.25) directly as the outer radius R_{out} influences the radiation energy entering into the system. The effect appears apparently as the 1D model is analyzed and calibrated in section 6.2.1 and 6.2.2. This shows that the 1D model is an over simplified model and one has to keep these limitation in mind.

6.2 Analysis and identification of the glass forming model with UQ Framework

As mentioned in chapter 1, the discrepancy between the real measurements and the numerical solution is a typical issue in modeling and simulation. This issue with respect to the glass forming model is illustrated by figures 6.11 and 6.12. Figure 6.11 shows sensor measurement data of a real production of preforms with $D < 100$ mm. The figure shows that there are uncertainties in the process. It can be seen that the diameter D has low uncertainties, while the temperatures have relatively large uncertainties.

The comparison of the measurements and the simulation results is shown in figure 6.12.

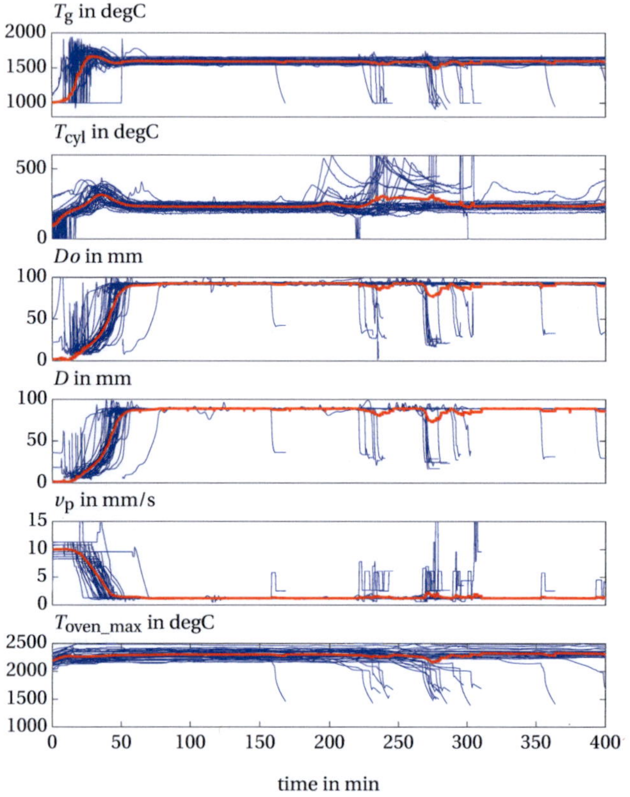

Figure 6.11: Measurement data from a production of preforms (blue line). The red lines represent the mean values.

Figure 6.12 presents three exemplary simulation results regarding to different parameter settings compared with the stationary measurement values from figure 6.11. It shows that the simulation results with the current parameter settings do not agree with the measurement data. Due to the nonlinearity of the model, it is very delicate to calibrate the model parameters in order that the model responds correspond to the real measurements.

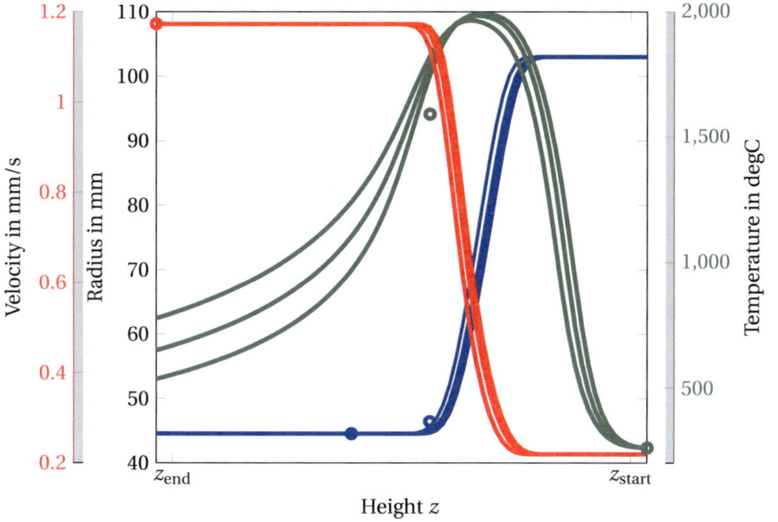

Figure 6.12: Solution of 1D model of glass rod regarding to different parameter settings (blue: radius, green: temperature, red: velocity). The circles illustrate the stationary measurements from figure 6.11.

In this section, the application of the framework of analysis and identification with UQ to the glass forming process model is presented to solve this disagreement. As mentioned in chapter 2, the incomplete knowledge in the modeling stage could lead to uncertainties in the model. Based on the Bayesian perspective, the uncertainties can be derived from the modeler's knowledge state. In the course of this thesis, all the uncertainties in the modeling are considered in form of uncertain parameters represented by random variables (RVs). In this application, the glass forming process model described previously is considered with the parameters represented by the RVs with some known probability distributions.

A related study is proposed by Mawardi in [Maw08]. The study shows the numerical simulation of an optical fiber drawing process under uncertain parameters. The considered model in his study is the 1D model of the glass fiber drawing. The propagation of the uncertainties of the input parameters to the model response uncertainty is computed based on a sampling-based method, namely *Latin Hypercube Sampling*.

Compared to the study in [Maw08], the considered model in this thesis requires more computational effort per simulation, that could not be feasible for a sampling-based method to calculate the pdf of the system response. The gPC expansion for the uncertainty propagation and the sensitivity analysis is applied to facilitate the computation. Moreover, the gPC approximation can be used as a surrogate model in the parameter calibration procedure, as discussed in section 5.3. In this thesis, the 1D and 2D glass forming model is analyzed and identified with the developed framework.

6.2.1 Analysis of 1D glass forming process model

In the considered glass forming model, there are in total about 30 concerned parameters. In order to understand the behavior of the system, the influence of each parameter should be found out by sensitivity analysis. In this section, the system analysis of the considered 1D glass forming model is separated into two studies, namely the parameters of the oven model and the parameters of the material model. The parameters of the oven model including the emissivity ϵ and the parameters of the material model are arranged in vector form as:

$$\boldsymbol{q}_{\text{oven}} = \left[T_{\text{oven\_max}}, T_{\text{amb}}, a_{\text{above}}, b_{\text{above}}, a_{\text{below}}, b_{\text{below}}, \epsilon\right]^T$$
$$\boldsymbol{q}_{\text{mat}} = \left[\lambda_1, \lambda_2, \lambda_3, c_{p_1}, c_{p_2}, c_{p_3}, \eta_1, \eta_2, \eta_3, \eta_4\right]^T$$

In the studies, these 17 parameters are considered to be uncertain. The parameters are assumed to be independent uniformly distributed. The uncertain range of the parameters is given by $\boldsymbol{q}_0 \pm \Delta\boldsymbol{q}$, where $\boldsymbol{q}_0$ is the nominal deterministic value of the parameters. The $\Delta\boldsymbol{q}$ is given usually in percentage based on the assumptions from the knowledge state of the model user. Figures 6.27a - 6.30a demonstrate the uncertainty of the parameters of the oven model and the material models.

The solution of the PDEs (6.23) - (6.25) are the temperature, velocity and the radius (calculated from the cross section area A) as functions of position along the z-axis as revealed in figure 6.7. According to the sensor system of the real production, the following quantities are considered as system responses of the 1D model.

- The *shrinkage* is the difference of the diameters between the two sensor positions $\Delta D := D(z_{s1}) - D(z_{s2}) = 2(R(z_{s1}) - R(z_{s2}))$.
- The temperature of the glass at the sensor position z_{s1}, called *glass temperature* $T_g = T(z_{s1})$.

The gPC expansions of the system responses are constructed for both cases, namely oven model and material model, regarding to the assumed uncertain parameters in each model. The gPC approximation is constructed with the procedure described in chapter 3 by using the software developed in the framework. As a consequence of the uniform distributions, the Legendre polynomials are used as orthogonal polynomial basis functions. The polynomial order P is chosen to be 4 according to the modeler's experience. For the reason that the considered model is implemented with commercial software and the solver code is difficult to access, the NISP method is used to calculate the coefficients of the gPC. Using the NISP approach, the polynomial coefficients are determined by the sparse grid quadrature (see section 3.2.4). With the FEM model described in section 6.1 the system responses at the sparse grid quadrature nodes are computed. The coefficients and the orthogonal polynomial constitute the gPC approximation model for the system responses mentioned above.

Uncertainty propagation

As discussed in chapter 3, the pdf of the system responses can be assessed from the gPC approximation model consisting of the orthogonal polynomial and the corresponding coefficients. The pdfs are approximated by using the gPC approximation as a surrogate model, which is computational cheaper than solving the full 1D model. The approximated pdfs of the two system responses of the 1D glass forming model T_g and ΔD are shown in figure 6.13 and 6.14. The pdfs indicate the probability of the system responses according to the assumed uncertain parameters. From both figures, it is indicated that the uncertainty in the oven model leads to the larger uncertainty of both model responses compared to those of the material model.

The uncertainties of the outer radius of the glass rod are presented in figure 6.15 for both models. The red line presents the mean of the outer radius, while the gray interval shows the uncertainty $\pm 2\sigma$ of the radius along the z-coordinate. The uncertainty interval is determined by the standard deviation value, which can be calculated directly from the gPC-coefficients.

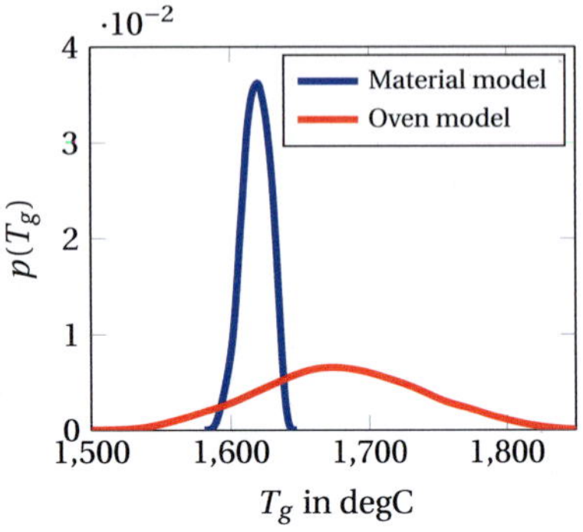

Figure 6.13: Estimated pdf of T_g

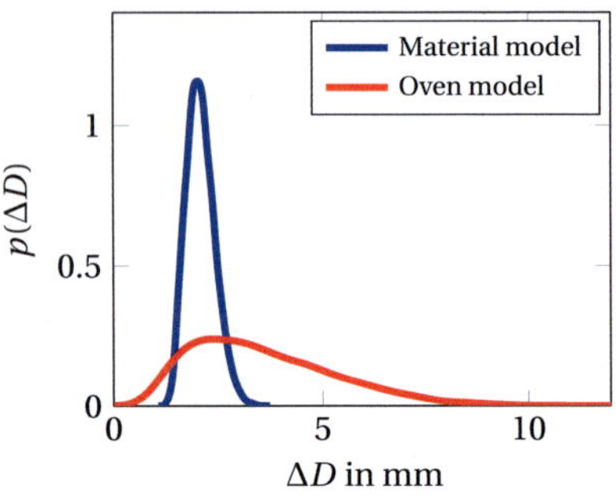

Figure 6.14: Estimated pdf of ΔD

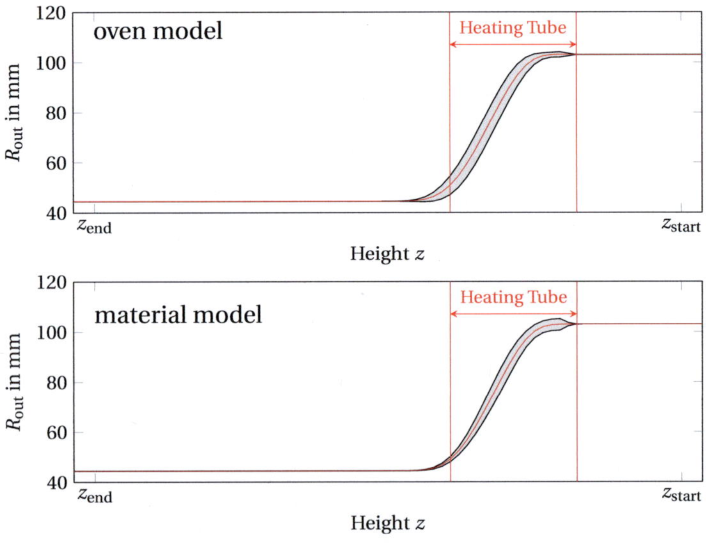

Figure 6.15: Uncertainty of the radius along the z-coordinate for the oven model and for the material model. (The values of z are intentionally suppressed due to a confidentiality obligations)

It can be seen that the uncertainty of the radius occurs solely in the deformation zone, but not including the boundary. This is comprehensible as the uncertain parameters in both studies do not act on the boundary conditions. These uncertainty intervals along the z-axis from both figures provide valuable information to the model user, where the uncertainties can arise. For example, figure 6.15 shows that the uncertainty of the material model has lower effect to the radius at the heating tube than the uncertainty of the oven model.

Figure 6.16 visualizes the pdf of the temperature T and along the z-axis of the oven model study. The temperature value at $z \approx z_{\text{start}}$ is almost deterministic because of the Dirichlet boundary condition . This appears in figure 6.16 as the pdf has a very high value at $z \approx z_{\text{start}} = 0.5$ m. A spatial slice of the pdf $p(T;z)$ at $z = z_{s1}$, shown by the magenta line in figure 6.16, constitutes exact the pdf $p(T_g)$ of the oven model shown in figure 6.13. This information delivers the uncertain range of the system responses regarding to the assumed uncertain parameters. In these studies it is obvious that the parameters of the oven model have more influence to both system responses (T_g,ΔD) than the parameters of the material model.

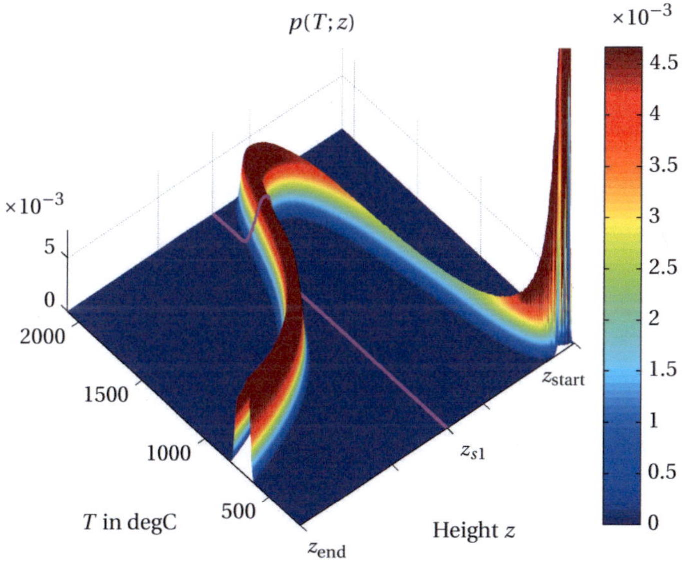

Figure 6.16: pdf of Temperature along the z-axis according to the uncertainty in the oven model. The magenta line presents a spatial slice of the pdf $p(T;z)$ at $z = z_{s1}$. (The values of z are intentionally suppressed due to a confidentiality obligations)

Sensitivity analysis

In the next step, the sensitivity analysis is applied to the system responses. A first study of sensitivity of the glass forming process using the finite difference method can be found in a previous publication [Ja11] of the author. In the course of this thesis, the sensitivity measures of all stochastic parameters are calculated from the gPC coefficients as discussed in section 4.4.

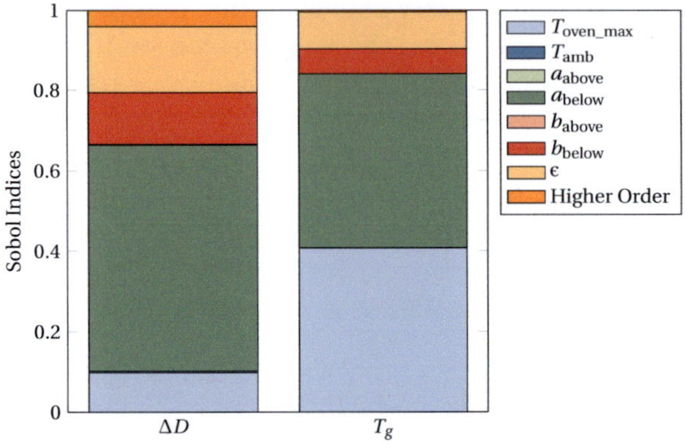

Figure 6.17: Sensitivities with respect to the oven parameters

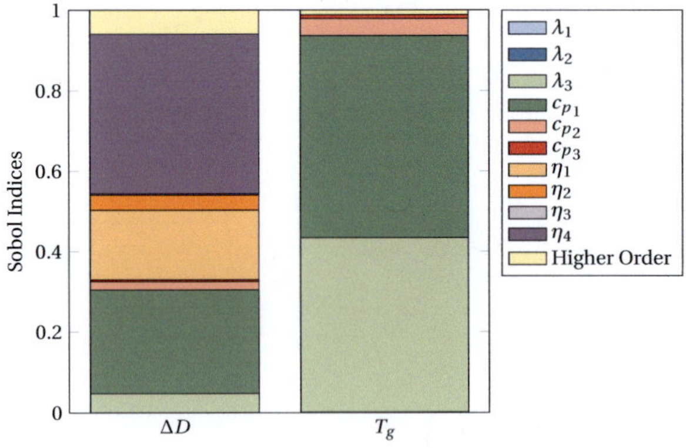

Figure 6.18: Sensitivities with respect to the material parameters

The sensitivity measures, i.e. Sobol Indices, of the parameters in the oven model and in the material model are illustrated in figure 6.17 and 6.18 respectively. The bar length indicates the relative influence of the parameters.

Both figures display that the parameters which have relatively large influence are $T_{\text{oven-max}}$, a_{below}, b_{below} and ϵ in the oven model and λ_3, c_{p_1}, η_1 and η_4 in the material model. The sensitivity information will be used for calibrating the parameters in the next section.

6.2.2 Parameter calibration of 1D glass forming process model

The parameter calibration of the glass forming process model by the Bayesian inverse problem approach without the gPC approximation is proposed in a previous publication [Ja13a] of the author. Due to the computational effort for solving the model, the calibration can be done only with a small number of parameters. In this section, the developed framework is applied to calibrate the model parameter with the help of the gPC approximation, which facilitates the calibration with a moderate number of parameters.

In the production process, the sensors provide measurements typically as time series as shown in figure 6.11. The production process consists basically of three main phases, namely start phase, stationary phase and end phase. In the course of the thesis, the system response in the stationary phase is considered because it mostly decides the quality of the end product. The measurement values used for the calibration can be obtained by the mean value at the stationary phase:

$$y_{ss} := \frac{1}{t_{\text{ss\_end}} - t_{\text{ss\_start}}} \int_{t_{\text{ss\_start}}}^{t_{\text{ss\_end}}} y(t)dt. \tag{6.32}$$

Apart from the measurements of T_g and ΔD, the following process quantities are taken into account for the calibration:

- the diameter of the glass cylinder D_{cyl},
- the cylinder temperature T_{cyl},

- the maximal oven temperature $T_{\text{oven\_max}}$, and
- the mass throughput $MPpc$, the process specific quantity related to the feeding speed v_{feed} and the pulling speed v_{pull}.

The relation of the stationary measurement data is shown by scatter plots in figure 6.19. It reveals that the measurements data are hardly correlated to each others.

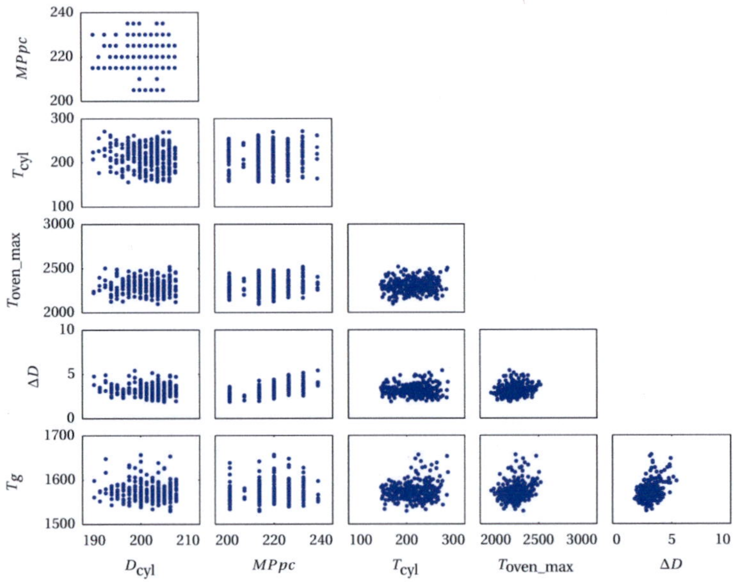

Figure 6.19: Scatter plots of the stationary measurements from selected productions (The units of the data are intentionally suppressed due to the confidentiality obligations.)

The parameters for calibration are chosen from the knowledge about the sensitivity of the parameters of the oven model and the material model. The following parameters are considered.

$$
\begin{aligned}
\boldsymbol{q} &= \left[\boldsymbol{q}_{\text{process}};\ \boldsymbol{q}_{\text{calibrate}}\right] \\
\boldsymbol{q}_{\text{process}} &= \left[D_z, MPpc, T_{\text{cyl}}\right]^T \\
\boldsymbol{q}_{\text{calibrate}} &= \left[T_{\text{oven\_max}}, \epsilon, a_{\text{below}}, b_{\text{below}}, c_{p_1}, \lambda_3, \eta_1, \eta_4\right]^T
\end{aligned}
$$

The study model is also considered in the form:

$$
\boldsymbol{y}^M = \begin{pmatrix} \Delta D \\ T_g \end{pmatrix} = \mathcal{M}\left(\boldsymbol{q}_{\text{process}}, \boldsymbol{q}_{\text{calibrate}}\right) \tag{6.33}
$$

The measurement $\boldsymbol{y}_{\text{ss}}$ is modeled by the simulation result with additive independent Gaussian white noise $\boldsymbol{w}(\omega)$, whose variance can be determined from the measurement data directly.

$$
\boldsymbol{y}_{\text{ss}} = \boldsymbol{y}^M(\boldsymbol{q}) + \boldsymbol{w} \tag{6.34}
$$

Based on the assumed pdfs of the parameters, the gPC approximation is constructed. The gPC approximation can be used to analyze the system and can be employed as a surrogate model in the parameter calibration.

$$
\boldsymbol{y}^M \approx \boldsymbol{y}^{gPC}\left(\boldsymbol{q}_{\text{process}}, \boldsymbol{q}_{\text{calibrate}}\right) \tag{6.35}
$$

The diameter of the preform in the real production is varied. In this study, only the productions of the preform with some certain diameter D are considered. The uncertainty propagation analysis based on the assumed inputs provides the results shown in figure 6.20 - 6.23. Figures 6.20 and 6.21 illustrate the uncertainty of the temperature and the radius along the z-coordinate. The sensitivity analysis is applied to the model and the Sobol indices of the two system responses w.r.t. the parameters are illustrated in figure 6.24.

The probability density functions of the system responses corresponding to the two sensor values are presented in figure 6.22 and 6.23. The measurements from the productions of the considered preforms are marked in the both figures. Both figures show that the assumed uncertainties of the parameters lead to the larger uncertainty of system responses than the fluctuation of the real measurements. The discrepancy between the system responses pdfs and the measurements suggest that the assumption of input parameter should be changed or rather calibrated.

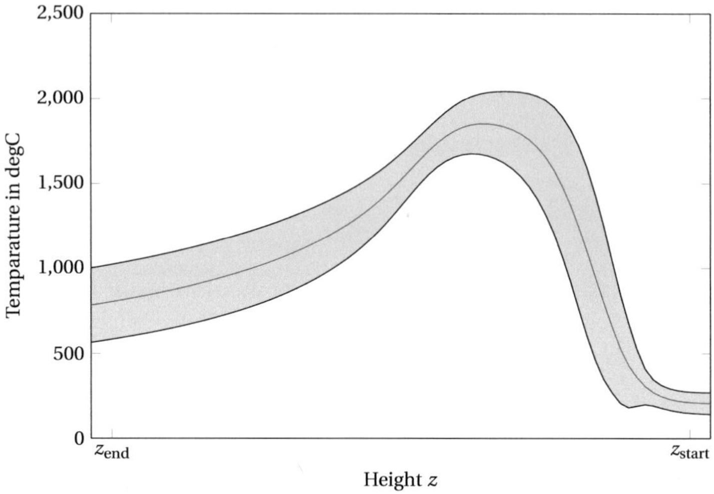

Figure 6.20: Temperature along the z-axis and its 2σ uncertainty interval (red line is the mean value)

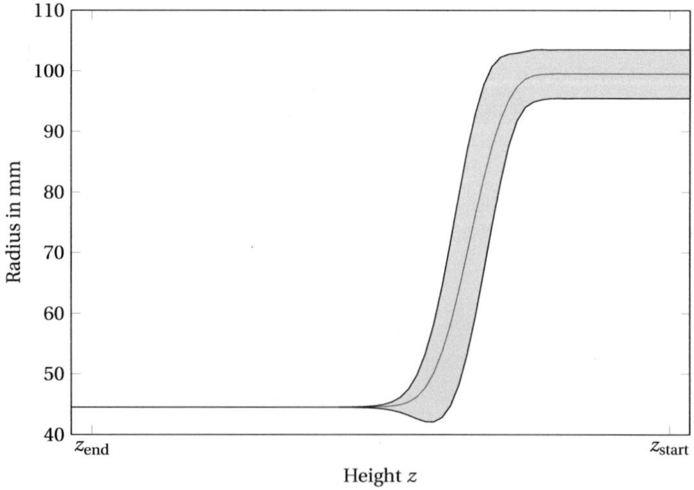

Figure 6.21: Radius along the z-axis and its 2σ uncertainty interval (red line is the mean value)

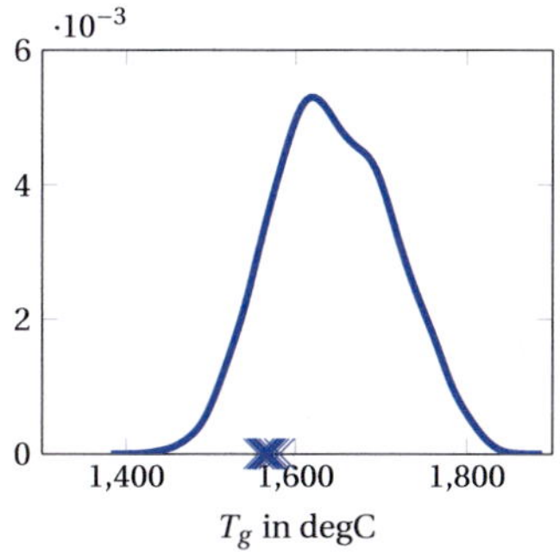

Figure 6.22: pdf of the T_g

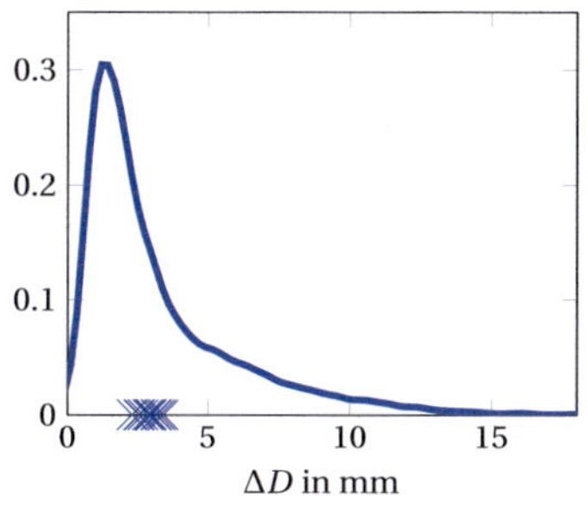

Figure 6.23: pdf of the ΔD

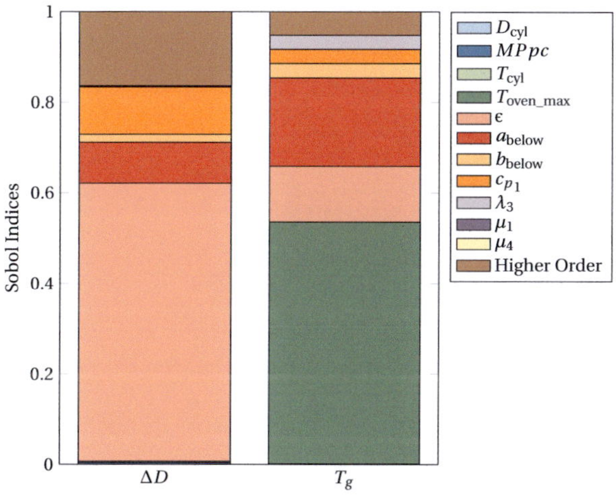

Figure 6.24: Sensitivity w.r.t. the parameters in the calibration model

In the Bayesian statistic inverse formulation, the input parameters $\boldsymbol{q}$, the measurement $\boldsymbol{y}$ and the model responses $\boldsymbol{y}^M$ are considered as the random variables. The parameter vector $\boldsymbol{q}_{\text{calibrate}}$ is calibrated using the Bayesian formulation as described in chapter 5.

In this section the batch method is applied to assess the posterior distribution. The computation of the posterior distribution by the MCMC method in this thesis is based on the implementation of *Delayed Rejection Adaptive Metropolis (DRAM)* by Haario [Haa06]. The estimated posterior pdfs of the calibrated parameters are presented in figure 6.25. The correlation of the parameters are illustrated by the scatter plots in figure 6.26. From the figure, one can observe the correlation between the parameters e.g. $T_{\text{oven\_max}}$, a_{below} and ϵ.

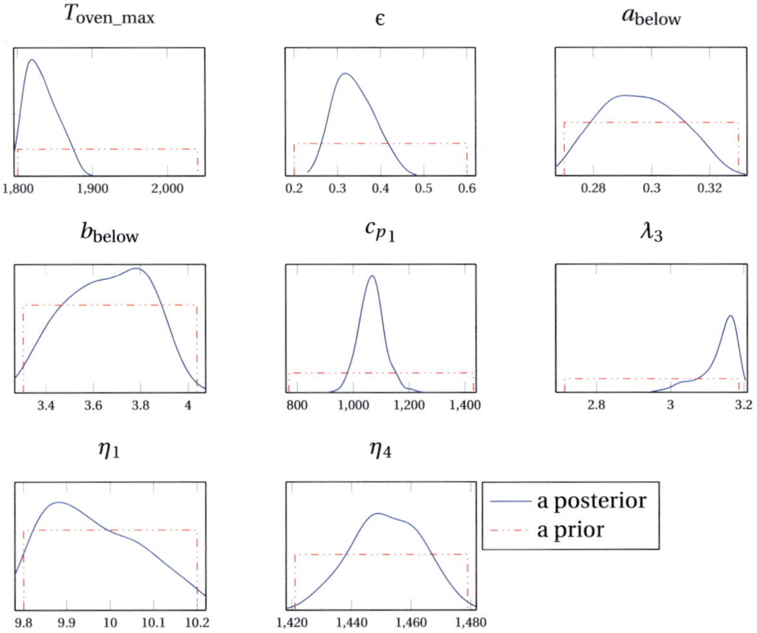

Figure 6.25: Estimated posterior pdf of the calibrated parameters (The units of the data are suppressed due to the confidentiality obligations)

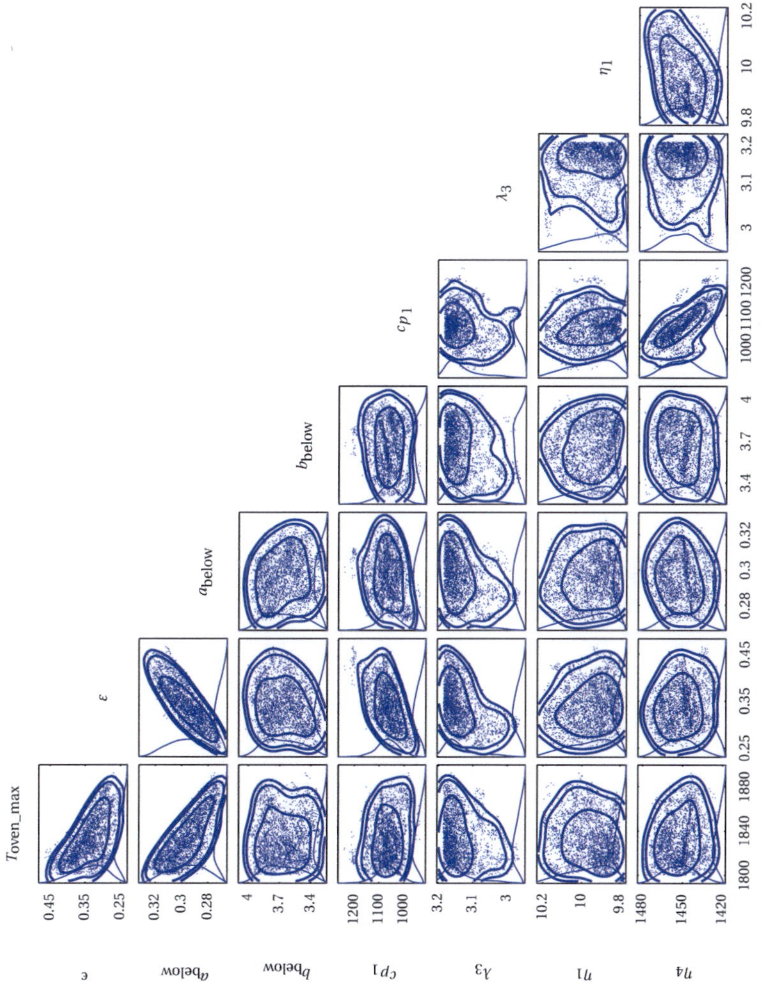

Figure 6.26: Scatter plots of the posterior distributions

The distribution changes of the parameters compared to their prior distribution can be observed in figure 6.25. These changes are also corresponding to the Sobol indices shown in figure 6.24 except for the parameter a_{below}. The parameter a_{below} does not change much despite its relatively high influence. This is due to the correlations existing between the parameter a_{below} and the parameter ϵ and $T_{\text{oven\_max}}$, which have higher influence than the parameter a_{below}.

The values of calibrated parameters $T_{\text{oven\_max}}$ and ϵ are relatively low compared to the physical meaningful parameter. As mentioned previously, in the parameter calibration the calibrated parameters could have little or no physical meaning. From the fact that the 1D model provides too low shrinkage value, as discussed in section 6.1.4, the Bayesian calibration provides the best possible parameter set regarding to the current model. This results in the calibrated parameters $T_{\text{oven\_max}}$ and ϵ, which are the most sensitive parameters as revealed in figure 6.24.

The difference between the prior and the posterior distribution of the oven model and the material model can be seen from figures 6.27 to 6.30. Figures 6.27a - 6.30a show the prior distributions, while figures 6.27b - 6.30b present the posterior distribution after the calibration. The most distinct changes happen for the parameters $T_{\text{oven\_max}}$, λ_3 and c_{p_1}. The distribution changes can be seen apparently in figures 6.27, 6.28 and 6.29.

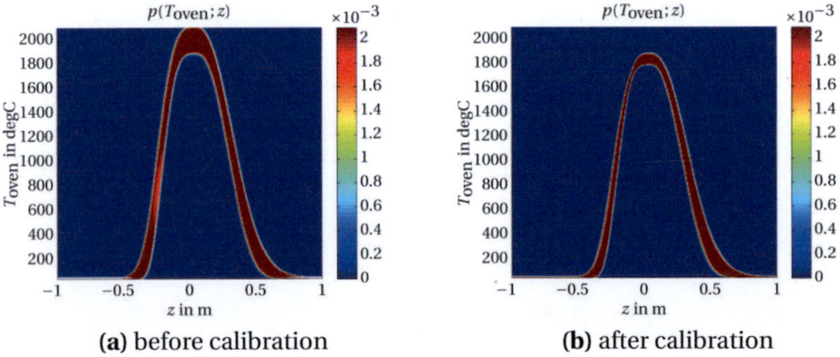

(a) before calibration

(b) after calibration

Figure 6.27: Distributions of the oven temperature

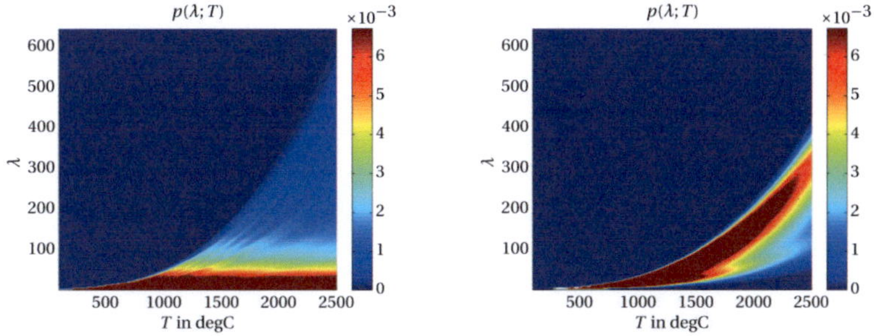

(a) before calibration **(b)** after calibration

Figure 6.28: Distributions of the heat conductivity

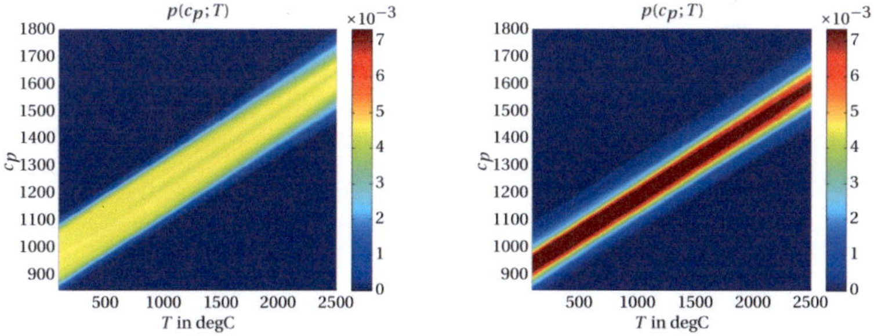

(a) before calibration **(b)** after calibration

Figure 6.29: Distributions of the heat capacity

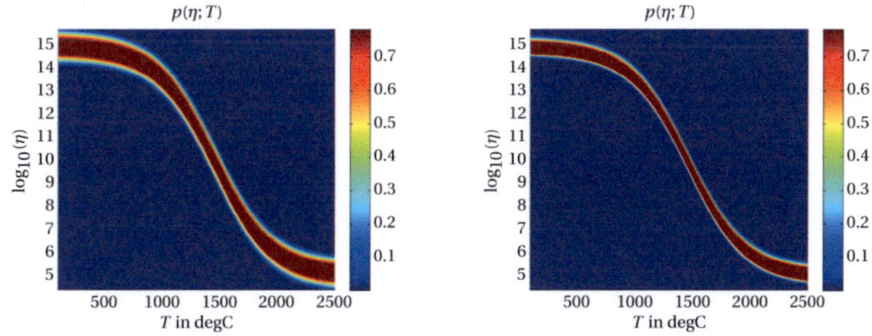

(a) before calibration **(b)** after calibration

Figure 6.30: Distributions of the viscosity

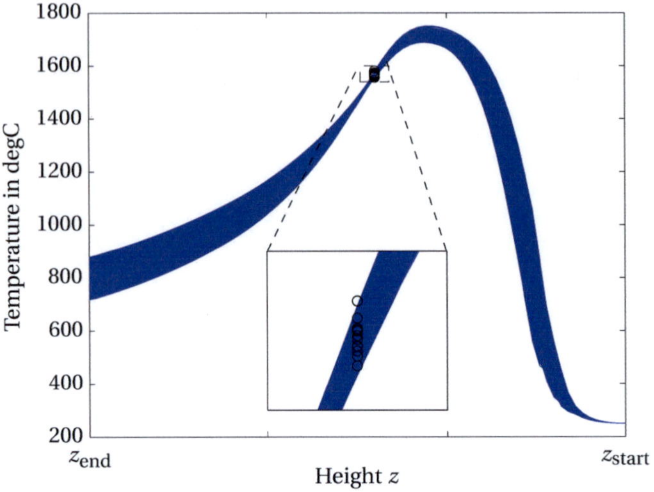

Figure 6.31: Distribution of the temperature along the z-coordinate including the T_g measurements (black circles)

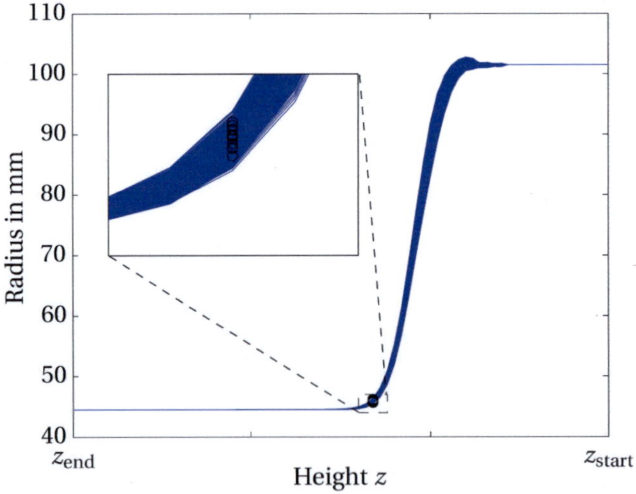

Figure 6.32: Distribution of the radius along the z-coordinate including the measurements (black circles)

By evaluating the model with the posterior pdfs, the temperature and the radius pdf along the z-coordinate result as shown in figure 6.31 and figure 6.32 respectively. Compared to the distributions before calibration shown in figure 6.20 and 6.21, the distributions with the calibrated parameter are more concentrated. In the magnified view of both figures, it is obvious that the curves go through the measurements, which are indicated by the black circles. This can be interpreted that the distributions of the calibrated parameters provide consistent results. The calibrated parameters yield system responses that correspond with the given measurement data. However, the model with the calibrated parameters could lose the physical interpretation, if the underlying model does not reflect the process with the desired accuracy, which is the case of the simplified 1D Trouton model.

Additionally to quantitative information about the calibrated parameters, one achieves the knowledge from system analysis and identification about the system as well. The effects of the parameter spreading through the spatial space can be derived from the uncertainty propagation of different parameter settings, as shown in e.g. figure 6.15 and 6.16. The intrinsic relation between the parameters can be obtained from the sensitivity analysis (Fig. 6.24) as well as the posterior distributions (Fig. 6.26). From this application of the framework to the 1D model, one can also realize that the assumed radiation energy determined by the oven model seems to be too high for the 1D model.

6.2.3 Analysis of 2D glass forming process model

As the second numerical study, the analysis and parameter calibration of the 2D model of the preform production process is considered. Due the radial dependence of the variables in the 2D model, which is discussed in section 6.1.4, the importance of the parameters the in 2D model is expected to be different from the 1D model.

The parameter terms which are related directly to the temperature such as c_{p_2} and λ_2 should be more interesting. Due to the computational effort of the 2D model, the study is restricted to three parameters. These parameters are two material parameters, namely, c_{p2} and λ_2 from equation (6.14) and (6.15) respectively, and one parameter of the oven temperature profile T_{o1}, as the parameters a_{below} and b_{below} are not physically representative and can not be interpreted in an easy way. The parameter T_{o1} is the oven temperature at the end of the oven and is related directly to the oven temperature T_{oven}. Together with the other specified process parameters such as T_{amb} and $T_{\text{oven\_max}}$, the parameters a_{below} and b_{below} are specified by a curve fitting method.

The solutions of PDEs (6.1) - (6.3) are the temperatures, and the velocity distribution in the 2D and the geometry of the deformed glass. The following quantities are considered as the system responses of the 2D model.

- The shrinkage $\Delta D = D(z_{s1}) - D(z_{s2}) = 2(R(z_{s1}) - R(z_{s2}))$. (see Fig. 6.10)
- The glass temperature $T_g = T(z_{s1})$.
- The viscous force at the position z_{s1}, which is defined as:

$$F_{\text{vis}}(z) := \int 2\eta(T(r,z)) \frac{\partial v_z(r,z)}{\partial z} 2\pi r\, dr \tag{6.36}$$

  $F_{\text{vis}}(z_{g1})$ corresponds to the measured pulling force F_{pull} of the production process.

These three system responses can be obtained by evaluating the solution of the PDEs resulted by the FEM calculation at the corresponding z-coordinate. The studied system is defined by these three input parameters and these three system responses as

$$\boldsymbol{y}^M = \begin{pmatrix} \Delta D \\ T_g \\ F_{\text{vis}} \end{pmatrix} = \mathcal{M}\left(T_{o1}, c_{p_2}, \lambda_2\right). \tag{6.37}$$

As in the 1D-study, the uncertainties of the the parameters is given by the model user in the form $\boldsymbol{q}_0 \pm \Delta \boldsymbol{q}$. The gPC approximation model with the polynomial order $P = 7$ is constructed based on the assumed uncertain parameters. The probability distribution of the three system responses are visualized by histograms and scatter plots in figure 6.33. The histograms show the probability range of the system response regarding the assumed input parameters, while the scatter plots reveal the dependency of the three system responses.

As sensitivity measures, the Sobol indices are calculated from the gPC coefficients. The Sobol indices of the three system responses are illustrated in figure 6.34. It is obvious that the parameter λ_2 has the most influence to the three system responses.

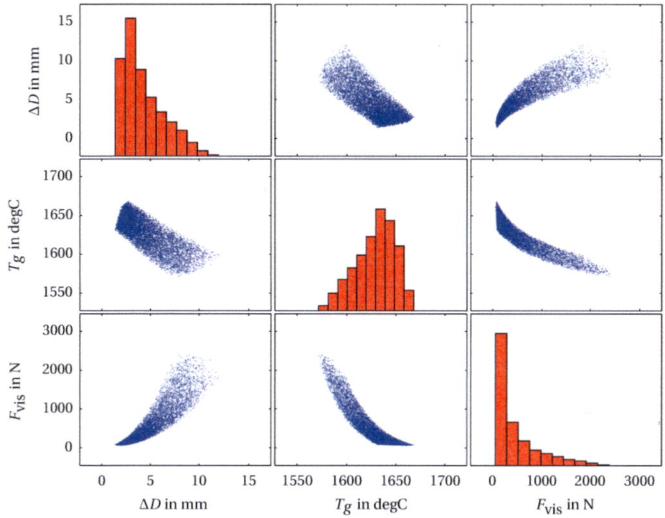

Figure 6.33: gPC approximated histograms and scatter plots of the 2D model responses regarding the assumed uncertainties of the parameters

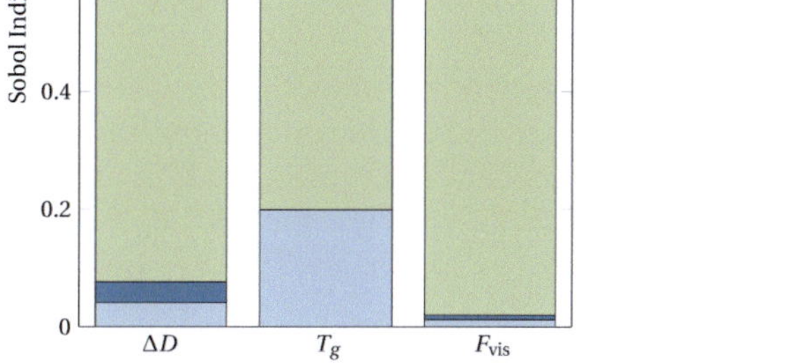

Figure 6.34: Sensitivity of the parameters in the 2D study

By using the gPC as a surrogate model, the system responses can be approximated. The result of approximated system responses subject to the parameter λ_2 by considering the parameters T_{o1} and c_{p_2} as deterministic nominal value is shown in figure 6.35. The figure shows a very good agreement between the gPC approximation and the original 2D FEM model.

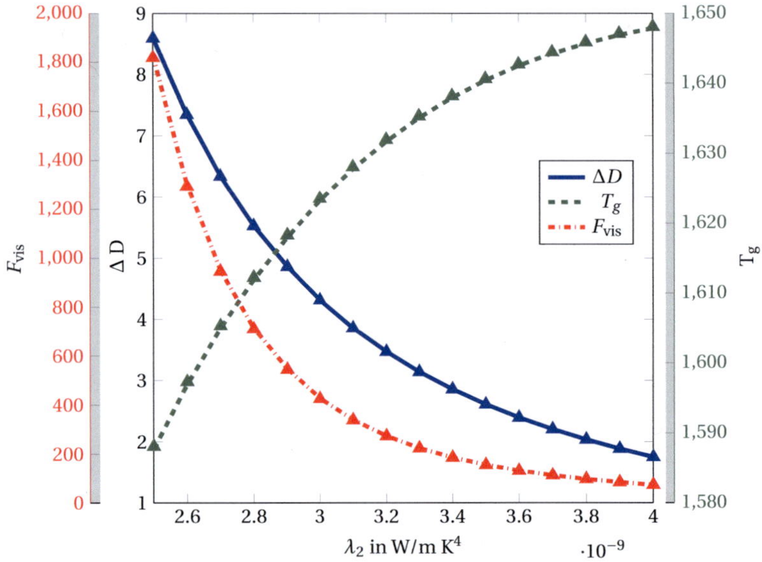

Figure 6.35: System responses (Δ) and its gPC approximation (line) of the 2D model subject to the parameter λ_2 by setting the parameter T_{o1} and c_{p_2} to fixed values.

Local sensitivity analysis is applied by computing the derivative of the model responses as described in section 4.2. As an example, the derivative of the system responses with respect to the parameter λ_2 is calculated and the result is presented in figure 6.36. The local sensitivities reveal the influence of the parameter λ_2 to all system responses. This information is very useful for the model user to develop and improve the model. As discussed in section 4.2, the approximation of the derivative works only in case that the PDEs solutions with respect to the parameter space are smooth enough.

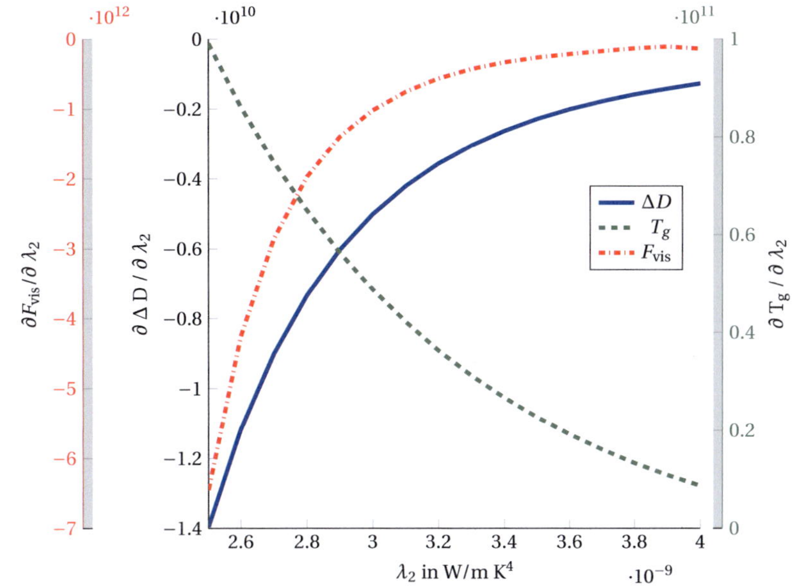

Figure 6.36: Derivative with respect to the parameter λ_2 based on gPC approximation of the system responses of the 2D model

6.2.4 Parameter calibration of 2D glass forming process model with Bayesian update

One goal of the study is to find the value of $\boldsymbol{q} = \left[T_{o1}, c_{p_2}, \lambda_2\right]^T$, which gives the computer model prediction $\boldsymbol{y}^M$ closest to the given measurement data $\boldsymbol{y}$. In this section, the recursive Bayesian update is applied to compute the posterior pdf of $\boldsymbol{q}$.

As discussed in chapter 5, the solving of the statistical inverse problem requires repeated computations of the model at different parameter settings. In order to avoid this extensive computation of the FEM, the gPC approximation is used as surrogate model for the forward model in the calculation of the likelihood function.

Representation by PCE

In the inverse uncertainty quantification framework, the input parameters $\boldsymbol{q}$, the measurement $\boldsymbol{y}$ and the model responses $\boldsymbol{y}^M$ are considered as the random variables (see chapter 5). Following section 5.2.1, the RVs can be represented by the PCE:

$$\boldsymbol{Q}(\omega) = \sum_{k=0}^{N_P} \boldsymbol{q}_k \boldsymbol{H}_k(\boldsymbol{\xi}(\omega)), \tag{6.38}$$

$$\boldsymbol{Y}(\omega) = \sum_{k=0}^{N_P} \boldsymbol{y}_k \boldsymbol{H}_k(\boldsymbol{\xi}(\omega)), \tag{6.39}$$

$$\boldsymbol{Y}^M(\omega) = \sum_{k=0}^{N_P} \boldsymbol{y}_k^M \boldsymbol{H}_k(\boldsymbol{\xi}(\omega)), \tag{6.40}$$

where $\boldsymbol{H}_k$ is multi-dimensional PC constructed from the one-dimensional Hermite Polynomials.

For computational purposes, in this study the expansion order of the PCE P is 5. With the dimensions of input parameters of 3, it yields $N_P + 1 = 56$ coefficients per outputs (see eq.(3.12)). These PCE representations of $\boldsymbol{Q}, \boldsymbol{Y}, \boldsymbol{Y}^M$ can be written in the matrix form

$$\mathbf{Q} = \left[\boldsymbol{q}_0 \cdots \boldsymbol{q}_k \cdots \boldsymbol{q}_{N_P}\right], \tag{6.41}$$

$$\mathbf{Y} = \left[\boldsymbol{y}_0 \cdots \boldsymbol{y}_k \cdots \boldsymbol{y}_{N_P}\right], \tag{6.42}$$

$$\mathbf{Y}^M = \left[\boldsymbol{y}_0^M \cdots \boldsymbol{y}_k^M \cdots \boldsymbol{y}_{N_P}^M\right]. \tag{6.43}$$

where $N_P = 55$ in this case. Given the means of the measurement μ and the corresponding standard deviation σ, the measurement $\mathbf{Y}$ is given by

$$\mathbf{Y} = \begin{bmatrix} \mu_{\Delta D} & \sigma_{\Delta D} & 0 & 0 & 0 & \cdots & 0 \\ \mu_{T_g} & 0 & \sigma_{T_g} & 0 & 0 & \cdots & 0 \\ \mu_{F_{\text{vis}}} & 0 & 0 & \sigma_{F_{\text{vis}}} & 0 & \cdots & 0 \end{bmatrix}. \tag{6.44}$$

The prior distribution of the parameter vector $\boldsymbol{q}$ is assumed as an independent Gaussian distribution, i.e.

$$\mathbf{Q}^0 = \begin{bmatrix} \mu_{T_{o1}} & \sigma_{T_{o1}} & 0 & 0 & 0 & \cdots & 0 \\ \mu_{c_{p2}} & 0 & \sigma_{c_{p2}} & 0 & 0 & \cdots & 0 \\ \mu_{\lambda_2} & 0 & 0 & \sigma_{\lambda_2} & 0 & \cdots & 0 \end{bmatrix}. \tag{6.45}$$

Figures 6.37 and 6.38 present the scatter plots of the measurements $\boldsymbol{Y}$ and the prior distribution $\boldsymbol{Q}^0$. Starting from the matrices $\mathbf{Y}$ and $\mathbf{Q}^0$, the Bayesian update of the parameter $\boldsymbol{q}$ can be performed as described in section 5.2.1. The model output $\boldsymbol{Y}^M$ is a mapping of $\boldsymbol{Q}$. The PCE representation of $\boldsymbol{Q}$ is summation of polynomial functions of normal distributed RV $\boldsymbol{\xi}(\omega)$. Therefore, the coefficient of $\mathbf{Y}^M$ can be obtained by the NISP method, where the integral of the inner product is computed by means of the Gauss-Hermite quadrature.

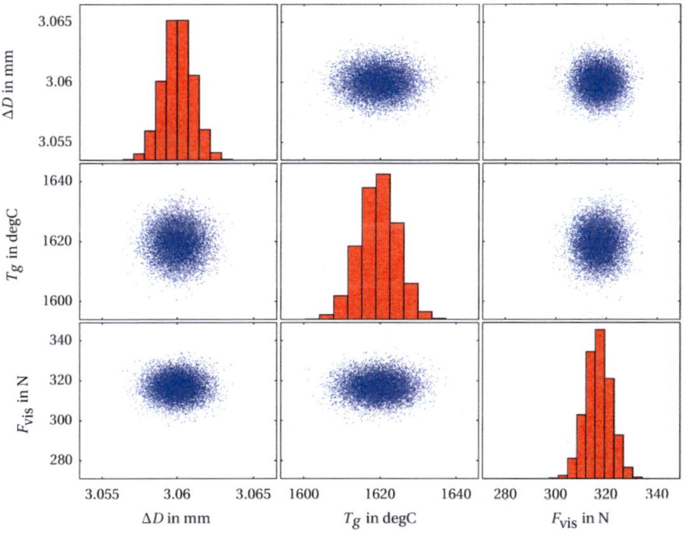

Figure 6.37: Distributions of the measurements $\boldsymbol{y}$

Bayesian update results

Figure 6.39 shows the parameter $\boldsymbol{Q}$ after the update in the first step. Compared to the prior distribution shown in figure 6.38, one can see that the pdfs of the parameter change significantly after the first step update. Although the prior distribution is set to be independent Gaussian, the pdfs after the first update shows dependency immediately. Moreover, the updated pdfs are not Gaussian anymore. This point reveals the advantage of Bayesian update with PCE over the original Kalman-Filter, which has the limitation to the Gaussian distribution as mentioned in section 5.2.1.

The Bayesian update is performed for $N_{\text{step}} = 20$ steps to show the convergence of the estimation and is illustrated in figure 6.40 and figure 6.41. Figure 6.40 shows the means of the estimations with the 2σ intervals. The mean values and standard deviations of the estimated parameters $\boldsymbol{q}$ can be determined directly by the PCE coefficients as discussed in chapter 3.

Figure 6.41 demonstrates the convergence of the estimation conclusively by the prediction error of the system responses. The relative error is determined by the following equation:

$$\epsilon = \frac{\left|\mathbb{E}(Y^M) - \mathbb{E}(Y)\right|}{|\mathbb{E}(Y)|}. \tag{6.46}$$

Figure 6.42 shows the final results of the parameter calibration after 20 updates. The strong dependency of the parameters, e.g. between c_{p2} and λ_2, is clearly visible. In order to evaluate the results, the parameter $\boldsymbol{Q}$ is also calibrated by using the MCMC method. The results of the calibration using the MCMC method are shown in figure 6.43.

The results from the PCE update method and MCMC method are quite similar. A small difference is that the posterior pdfs resulted by the MCMC methods is more distributed than by the PCE update method. This difference in this study could be caused by the slow convergence rate of the MCMC compared to the PCE update method. This is also the definitive advantage of the PCE update method compared to batch methods. However, results from the PCE update method could be differ from the batch method with sufficient number of samples for nonlinear model as the PCE update is a linear method. The reader is referred to [Mat16b] for detail about the nonlinear version of the PCE update method.

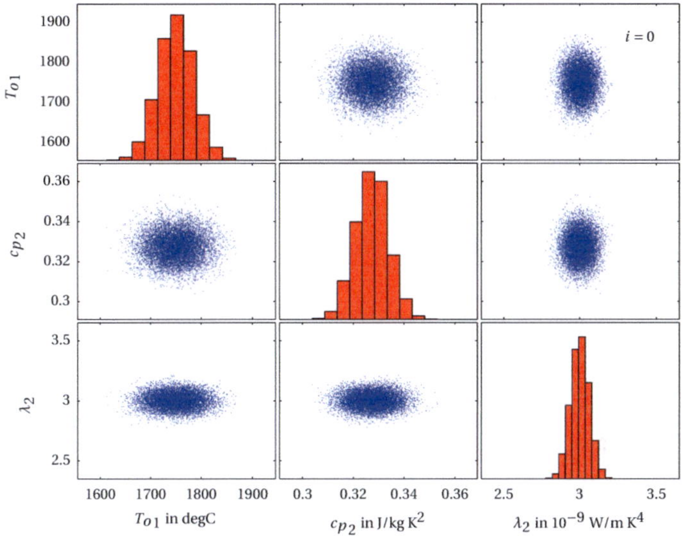

Figure 6.38: Prior distributions of parameters $\boldsymbol{q}$

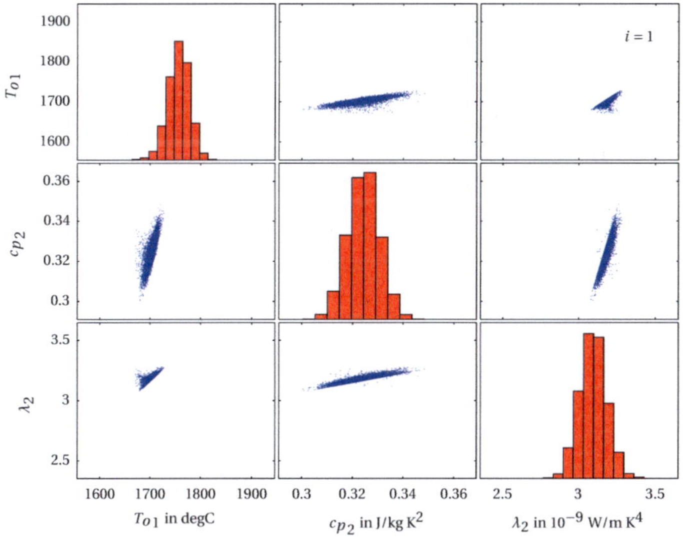

Figure 6.39: Distributions of parameters $\boldsymbol{q}$ after the first update step

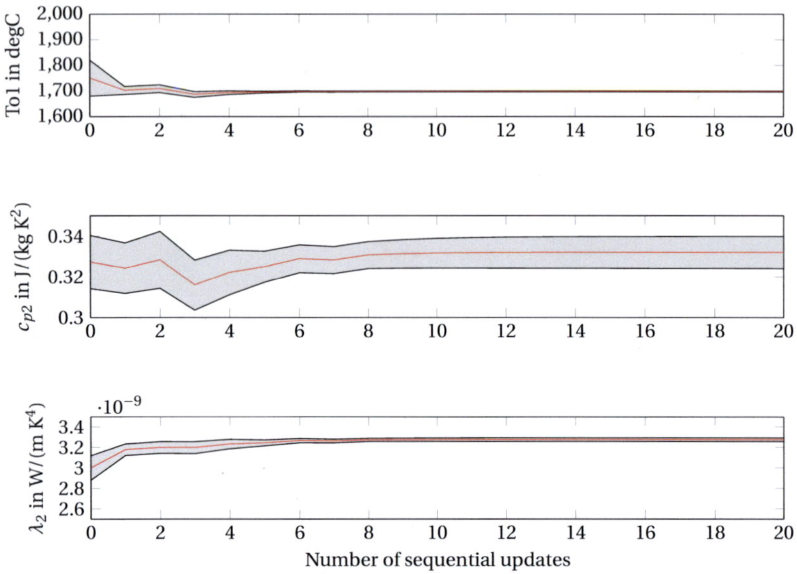

Figure 6.40: Mean values and 2σ intervals of the estimated parameters

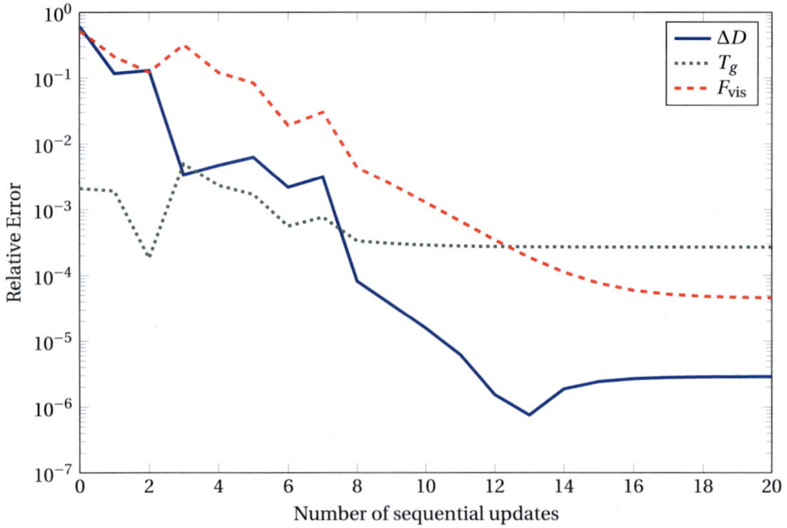

Figure 6.41: Error of the model prediction using the estimated parameters

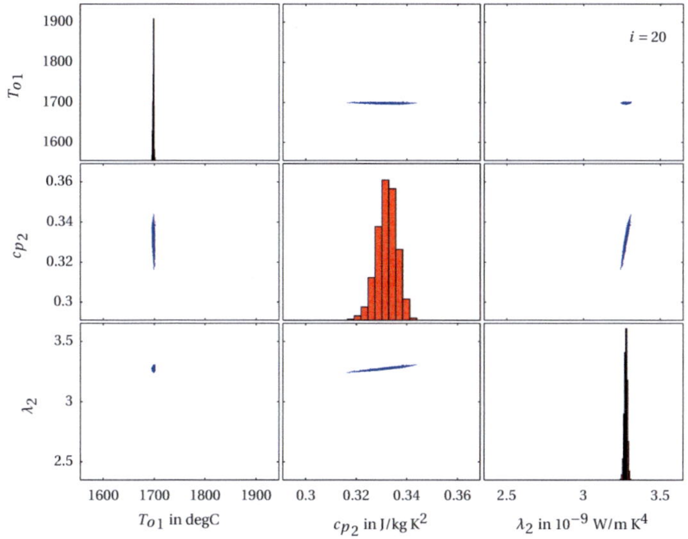

Figure 6.42: Distributions of the parameters $\boldsymbol{q}$ after update 20 step

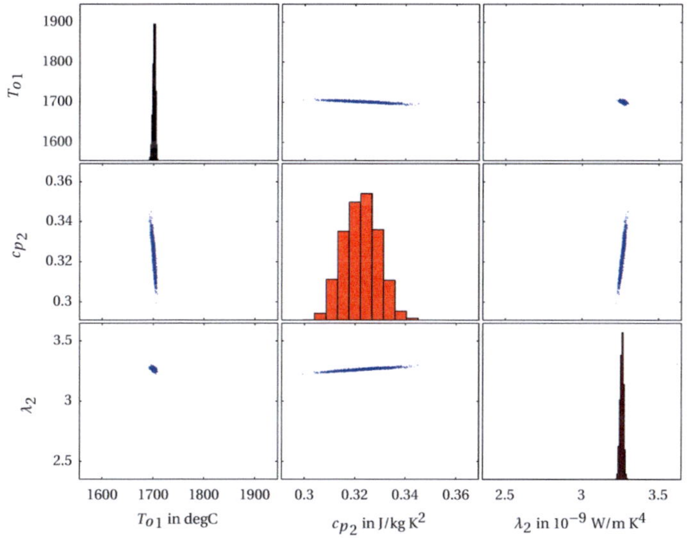

Figure 6.43: Distributions of the parameters $\boldsymbol{q}$ computed by MCMC

It can be acknowledged from Fig. 6.40, 6.42, and 6.43, that the estimation of the parameter c_{p_2} still has a large variance compared to the other two parameters. This is because the parameter c_{p2} of the specific heat capacity has the smallest influence to the three system responses compared to the other two parameters as the sensitivity analysis finds out (cf. Fig. 6.34).

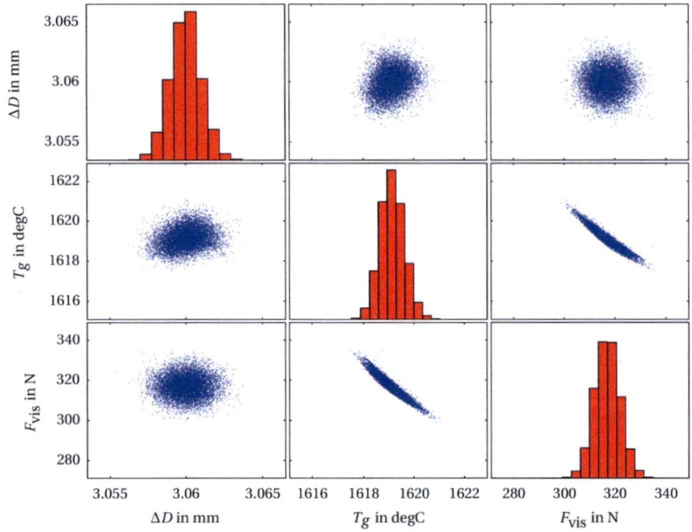

Figure 6.44: Distributions of the model prediction $\boldsymbol{y}^M$

The convergence of the error shown in figure 6.41 can be construed at once as the limit of the calibration by using only the three input parameters. It shows that the other parameters should be taken into account to obtain better results with less errors. Nevertheless the relative error of ca. 7×10^{-3}, which equates to less than 1 K error, suffices more than enough with respect to the considered application.

The pdf of $\boldsymbol{Y}^M$ based on the calibrated parameters $\boldsymbol{Q}$ after the 20 updates is presented in figure 6.44. Compared to the pdfs of the given measurement $\boldsymbol{Y}$ shown in figure 6.37, the pdfs of $\boldsymbol{Y}^M$ give the similar mean values as the measurement $\boldsymbol{Y}$. However, the pdfs of $\boldsymbol{Y}^M$ does not show the independencies as it has been assumed. Figure 6.44 reveals some dependency of the system responses, especially between T_g and F_{vis}. This shows that these two system responses possess some relationship in this model. The reason for this difference is that the distribution of the measurement $\boldsymbol{Y}$ is due to the measurement noise $\boldsymbol{W}(\omega)$, whereas the distribution of the $\boldsymbol{Y}^M$ shows the systematic uncertainty resulting from the uncertainty in the parameter.

6.3 Conclusion

The model of a real-world industrial glass forming process is a complex distributed parameter system, which can be described by the system of nonlinear PDEs discussed in section 6.1. The parameters of the PDEs determine the behavior of the system responses. The framework of analysis and identification proposed in this thesis is applied to the simplified 1D model and more realistic, to the axisymmetric 2D model of the glass forming to analyze and identify the system with its uncertain parameters. All the computations are feasible thanks to the efficiency of the gPC approach. The combination of the uncertainty propagation, sensitivity analysis and the Bayesian parameter calibration provide new aspects of the modeling. Although, the calibration results of the 1D model cannot be exploited directly as the underlying model does not reflect accurately enough the real process, the knowledge that one gains from the UP, the SA and posterior probability distribution is still valuable. The knowledge obtained from the analysis and the identification is fundamental for improving the glass forming model and the process in the future.

7 Conclusion

7.1 Summary of contributions

The main goal of the thesis is to elucidate a systematic approach that allows one to gain an understanding of the behavior of nonlinear distributed parameter system (DPS) and reducing the discrepancy between the computational model predictions and the process measurements. A framework for analysis and identification of nonlinear DPSs using Bayesian approach is also developed in this thesis. Under the Bayesian approach, the state of knowledge is formulated into uncertainties. The uncertainties are represented by the random variables or random fields, whose probability distributions can be derived by the Maximum Entropy Principle. The modeling of uncertainties in DPS yields a system of partial differential equations (PDEs) with random variables (RVs). To solve the system of PDEs with RVs, the PDEs are discretized by appropriate methods. Various uncertainty quantification (UQ) approaches are employed to harvest the knowledge about the system, which could be used to improve the model further.

The UQ approaches are applied in this thesis to analyze and to identify the system. As the UQ approaches require extensive computation, the generalized polynomial chaos (gPC) is applied to reduce the computational effort. The efficiency comparison between the gPC method and conventional sampling-based methods is presented by means of numerical examples of uncertainty propagation. It shows that the gPC can provide accurate pdf approximations with much lower computational effort compared to the sampling-based methods.

Sensitivity analysis is used to quantify the relative contributions of the input parameters to the system responses. The exploitation of gPC for both local and global sensitivity analysis are also presented. The comparison of the computation of sensitivities with gPC to the conventional method is also demonstrated by means of numerical examples. It shows that the local sensitivities can be approximated by using the gPC as a surrogate model. The error of the approximation depends on the expansion order of gPC. The main advantage of the approach is that the method considers the system as a black-box and does not need any analytical solution to calculate the derivative, which is often not available for DPS. Furthermore, the global sensitivity measures, the Sobol Indices, can be effectively computed from the gPC coefficients.

The Bayesian statistic formulation of an inverse problem and computational methods, both batch and recursive, are discussed. The exploitation of polynomial chaos expansion (PCE), namely using as a surrogate model, and using in the linear Bayesian update are presented. The efficiencies of each method are demonstrated by means of numerical examples. The calibration methods with the help of the gPC approximation can provide the results very similar to the results without the approximation. This means that the gPC approximation can reduce the computational effort of the Bayesian inverse computation.

The framework proposed in this thesis is applied to the real-world industry process, namely the glass forming process. The framework is applied to the 1D and 2D glass forming model to analyze in order to harvest knowledge about the process and to calibrate the parameters of the model. It is also shown that the framework provides the parameter estimates which lead to a very good fit of the system model responses to the given measurement data. However, the parameters could not represent physical reality in the 1D model case, as the 1D model is modeled with an oversimplified assumption in this case.

7.2 Outlook to future work

There are a number of extensions which could advance the framework developed in this thesis. In this thesis, only the epistemic uncertainties due to the incomplete knowledge during the construction of a conceptual model are considered. The computational model is assumed to be error free and uncertainties due to measurement noise is also disregarded in general. However, the difference between the uncertainties due to the uncertain parameters and uncertainties due to measurement noises can arise as presented in section 6.2.4. The numerical error could also be significant, if one employs a too roughly discretized model. Both types of uncertainties could have different behaviors. The extension of the framework by integrating these two types of uncertainties into the framework could be an attractive future work.

In this thesis, only a low to moderate dimensional random variables are considered. Moreover, the independency assumption of each RV is also a requirement in order to calculate the gPC coefficients with the NISP method. The framework could be improved by pursuing a large number of dependent random variables. However, there are two main challenges for this issue. Firstly, the NISP method is not an optimal method to compute the gPC coefficicents anymore as the node number of the sparse grid quadrature would likely to exceed with the dimension number of RVs. Moreover, the NISP method is only applicable for independent RVs. Second, the number of expansion coefficients would be overwhelming as well with a too high stochastic dimension d.

For this extension the computation algorithms should be adapted. In case of a large number of of *independent* RVs, the *adaptive sparse grid* can be applied to obtain the similar accuracy with a lower number of quadrature nodes. An approach of adaptive sparse grid is discussed for example in [Ger98, LM10].

Another approach is to reduce the number of expansions terms. In some cases, some polynomial chaos terms have an insignificant influence to the system response, such that they could be neglected. An example of this approach is proposed by Blatman and Sudret, the number of expansion terms could be reduced by using a *hyperbolic truncation set*. The expansion with the hyperbolic truncation set is termed as *sparse polynomial chaos expansion* [Bla10, Bla08]. The algorithm to build up sparse polynomial chaos is explained in [Bla11].

The *least angle regression* [Efr04] could be applied to compute the coefficients. As it is a stochastic collocation method, the method is applicable for dependent RVs. The development in the direction of the *sparse polynomial chaos* and the integration of the *least angle regression* into the framework could be one interesting future works.

As discussed in chapter 3, non-intrusive approaches are preferable as one of the most important specification in engineering practice is to use existing deterministic solvers. However, there is still lack of theoretical study regarding the error estimation and convergence rate for the non-intrusive approaches. On the other hand, the stochastic Galerkin method ensures the orthogonality of the residue of the projection, but the modification of existing code is required. The idea of formulation the stochastic Galerkin in an non-intrusive way is presented and discussed with a simple example in [Gir14]. The non-intrusive Galerkin approach is one of the most interesting future works.

The gPC expansion has a fast convergence rate in case of smooth system responses regarding to the parameter space, but not for discontinuous responses. In case of discontinuous responses, other basis functions could be applied, e.g. *wavelet* [LM10] or Padé-Legendre [Cha09].

The gPC expansion is a suitable method as it matches the probability distributions commonly applied in engineering. A data-driven generalization is presented under the name *arbitrary Polynomial Chaos* (aPC) [Ola12]. The aPC generalize the polynomial chaos techniques in the direction of using arbitrary probability distributions from the data set. The investigation of this topic could also be one of the future works.

In the proposed framework, the linear Bayesian update is implemented as discussed in chapter 5. The non-linear Bayesian update and their formulation as *Gauss-Markov-Kalman filter*, proposed in [Mat16b, Mat16a], could be one of possible improvement of the parameter estimation approach.

Finally, the framework proposed in this thesis is devised for distributed parameter systems. Although, in this thesis only the application of the framework to the glass forming process and the neutron-diffusion example was done, the application of the framework, in principle could be applied to any distributed parameter systems.

Bibliography

[ame06] Guide for verification and validation in computational solid mechanics (2006), ASME, American Society of Mechanical Engineers, New York

[Ars14] ARSLAN, Erkan and CACUCI, Dan G.: Predictive modeling of liquid-sodium thermal–hydraulics experiments and computations. *Annals of Nuclear Energy* (2014), vol. 63:pp. 355–370

[Asc98] ASCHER, Uri M. and PETZOLD, Linda R.: *Computer methods for ordinary differential equations and differential-algebraic equations*, vol. 61, Siam (1998)

[Bab04] BABUSKA, Ivo; TEMPONE, Raúl and ZOURARIS, Georgios E.: Galerkin finite element approximations of stochastic elliptic partial differential equations. *SIAM Journal on Numerical Analysis* (2004), vol. 42(2):pp. 800–825

[Bad12] BADEA, Madalina C.; CACUCI, Dan G. and BADEA, Aurelian F.: Best-Estimate Predictions and Model Calibration for Reactor Thermal Hydraulics. *Nuclear science and engineering* (2012), vol. 172(1):pp. 1–19

[Ber79] BERNARDO, Jose M.: Reference posterior distributions for Bayesian inference. *Journal of the Royal Statistical Society. Series B (Methodological)* (1979):pp. 113–147

[Ber06] BERNARD, Thomas and EBRAHIMI MOGHADDAM, E.: Nonlinear model predictive control of a glass forming process based on a Finite Element model (2006), in: *2006 IEEE International Symposium on Intelligent Control*, pp. 960–965

[Bey99] BEYERER, Jürgen: *Verfahren zur quantitativen statistischen Bewertung von Zusatzwissen in der Meßtechnik*, VDI-Verlag (1999)

[Bla08] BLATMAN, Géraud and SUDRET, Bruno: Sparse polynomial chaos expansions and adaptive stochastic finite elements using a regression approach. *Comptes Rendus Mécanique* (2008), vol. 336(6):pp. 518–523

[Bla10] BLATMAN, Géraud and SUDRET, Bruno: An adaptive algorithm to build up sparse polynomial chaos expansions for stochastic finite element analysis. *Probabilistic Engineering Mechanics* (2010), vol. 25(2):pp. 183–197

[Bla11] BLATMAN, Géraud and SUDRET, Bruno: Adaptive sparse polynomial chaos expansion based on least angle regression. *Journal of Computational Physics* (2011), vol. 230(6):pp. 2345–2367

[Bou10] BOURINET, Jean-Marc: Ferum 4.1 user's guide. *Clermont-Ferrand, France: Institute Français de Mécanique Avancée (IFMA)* (2010)

[Buc88] BUCHER, Christian G.: Adaptive sampling-an iterative fast Monte Carlo procedure. *Structural Safety* (1988), vol. 5(2):pp. 119–126

[Bur] BURKARDT, John: Sparse Grid Mixed Growth Anisotropic Rules, URL `http://people.sc.fsu.edu/~jburkardt/cpp_src/sgmga/sgmga.html`, visited on 2016-02-12

[Cac03] CACUCI, Dan G.: *Sensitivity and uncertainty analysis, volume I: Theory*, Chapman & Hall/CRC, Boca Raton (2003)

[Cac05] CACUCI, Dan G.; IONESCU-BUJOR, Mihaela and NAVON, Ionel M.: *Sensitivity and uncertainty analysis, volume II: applications to large-scale systems*, Chapman & Hall/CRC, Boca Raton (2005)

[Cac14a] CACUCI, Dan G. and BADEA, Madalina C.: Predictive modeling of coupled multi-physics systems: II. Illustrative application to reactor physics. *Annals of Nuclear Energy* (2014), vol. 70:pp. 279–291

[Cac14b] CACUCI, Dan Gabriel: Predictive modeling of coupled multi-physics systems: I. Theory. *Annals of Nuclear Energy* (2014), vol. 70:pp. 266–278

[Cam47] CAMERON, Robert H. and MARTIN, William T.: The Orthogonal Development of Non-Linear Functionals in Series of Fourier-Hermite Functionals. *Annals of Mathematics* (1947), vol. 48(2):pp. p 385–392

[Cha09] CHANTRASMI, Tonkid; DOOSTAN, Alireza and IACCARINO, Gianluca: Padé–Legendre approximants for uncertainty analysis with discontinuous response surfaces. *Journal of Computational Physics* (2009), vol. 228(19):pp. 7159–7180

[Che03] CHEN, Zhe: Bayesian filtering: From Kalman filters to particle filters, and beyond. *Statistics* (2003), vol. 182(1):pp. 1–69

[Cho05] CHOUDHARY, Manoj K. and POTTER, R.M.: Heat transfer in glass-forming melts. *Properties of Glass Forming Melts* (2005):pp. 249–293

[Cho10] CHOUDHARY, Manoj K.; VENUTURUMILLI, Raj and HYRE, Matthew R.: Mathematical Modeling of Flow and Heat Transfer Phenomena in Glass Melting, Delivery, and Forming Processes. *International Journal of Applied Glass Science* (2010), vol. 1(2):pp. 188–214

[Cle60] CLENSHAW, Charles W. and CURTIS, Alan R.: A method for numerical integration on an automatic computer. *Numerische Mathematik* (1960), vol. 2(1):pp. 197–205

[Cou28] COURANT, Richard; FRIEDRICHS, Kurt and LEWY, Hans: Über die partiellen Differenzengleichungen der mathematischen Physik. *Mathematische Annalen* (1928), vol. 100(1):pp. 32–74

[Cre09] CRESTAUX, Thierry; LE MAÎTRE, Olivier and MARTINEZ, Jean-Marc: Polynomial chaos expansion for sensitivity analysis. *Reliability Engineering & System Safety* (2009), vol. 94(7):pp. 1161–1172

[DC95] DE COOMAN, Gert; RUAN, Da and KERRE, Etienne: Foundations and applications of possibility theory. *Proceedings of FAPT 95, World Scientific, Singapore, 328 biz* (1995)

[Don82] DONEA, Jean; GIULIANI, S. and HALLEUX, Jean-Pierre: An arbitrary Lagrangian-Eulerian finite element method for transient dynamic fluid-structure interactions. *Computer methods in applied mechanics and engineering* (1982), vol. 33(1-3):pp. 689–723

[dR08] DE ROCQUIGNY, Etienne; DEVICTOR, Nicolas and TARANTOLA, Stefano: *Uncertainty in industrial practice: a guide to quantitative uncertainty management*, John Wiley & Sons (2008)

[Efr04] EFRON, Bradley; HASTIE, Trevor; JOHNSTONE, Iain; TIBSHIRANI, Robert ET AL.: Least angle regression. *The Annals of statistics* (2004), vol. 32(2):pp. 407–499

[Eli00] ELISHAKOFF, Isaac: *Whys and hows in uncertainty modelling*, Springer (2000)

[Far11] FARINA, Angiolo; FASANO, Antonio and MIKELIĆ, Andro: Nonisothermal flow of molten glass: mathematical challenges and industrial questions, in: *Mathematical Models in the Manufacturing of Glass*, Springer (2011), pp. 173–224

[Fej33] FEJÉR, Leopold: Mechanische quadraturen mit positiven cotesschen zahlen. *Mathematische Zeitschrift* (1933), vol. 37(1):pp. 287–309

[Fie79] FIESSLER, Bernd; RACKWITZ, Rudiger and NEUMANN, Hans J.: Quadratic limit states in structural reliability. *Journal of the Engineering Mechanics Division* (1979), vol. 105(4):pp. 661–676

[Gan06] GANESAN, Sashikumaar: *Finite element methods on moving meshes for free surface and interface flows*, docupoint-Verl, Magdeburg (2006)

[Gar88] GARD, Thomas C.: *Introduction to stochastic differential equations*, M. Dekker (1988)

[Gel14] GELMAN, Andrew; CARLIN, John B.; STERN, Hal S. and RUBIN, Donald B.: *Bayesian data analysis*, vol. 2, Taylor & Francis (2014)

[Ger98] GERSTNER, Thomas and GRIEBEL, Michael: Numerical integration using sparse grids. *Numerical algorithms* (1998), vol. 18(3-4):pp. 209–232

[Gha03] GHANEM, Roger and SPANOS, Pol D.: *Stochastic finite elements: a spectral approach*, DoverPublications. com (2003)

[Gir14] GIRALDI, Loïc; LITVINENKO, Alexander; LIU, Dishi; MATTHIES, Hermann G. and NOUY, Anthony: To Be or Not to Be Intrusive? The Solution of Parametric and Stochastic Equations—the "Plain Vanilla" Galerkin Case. *SIAM Journal on Scientific Computing* (2014), vol. 36(6):pp. A2720–A2744

[Gre11] GREWAL, Mohinder S.: *Kalman filtering*, Springer (2011)

[Gro11] GROOT, J.A.W.M.; MATTHEIJ, Robert M. M. and LAEVSKY, K. Y.: Mathematical modelling of glass forming processes, in: *Mathematical Models in the Manufacturing of Glass*, Springer (2011), pp. 1–56

[Haa06] HAARIO, Heikki; LAINE, Marko; MIRA, Antonietta and SAKSMAN, Eero: DRAM: Efficient adaptive MCMC. *Statistics and Computing* (2006), vol. 16:pp. 339–354

[Had23] HADAMARD, Jacques: *Lectures on Cauchy's problem in linear partial differential equations, by Jacques Hadamard,..*, Yale University press, New Haven (1923)

[Han92] HANSEN, E: Global optimization using interval analysis. *Marcel Dekkar, New York* (1992)

[Has74] HASOFER, Abraham M. and LIND, Niels C.: Exact and invariant second-moment code format(for reliability analysis in multivariate problems). *American Society of Civil Engineers, Engineering Mechanics Division, Journal* (1974), vol. 100:pp. 111–121

[HD13] HAWKINS-DAARUD, Andrea; PRUDHOMME, Serge; VAN DER ZEE, Kristoffer G. and ODEN, J. Tinsley: Bayesian calibration, validation, and uncertainty quantification of diffuse interface models of tumor growth. *Journal of mathematical biology* (2013), vol. 67(6-7):pp. 1457–1485

[Hig10] HIGDON, David; HEITMANN, Katrin; LAWRENCE, Earl and HABIB, Sajad: Using the Bayesian framework to combine simulations and physical observations for statistical inference. *Large-Scale Inverse Problems and Quantification of Uncertainty* (2010):pp. 87–105

[Hig12] HIGDON, Dave; GATTIKER, James; WILLIAMS, Brian and RIGHTLEY, Maria: Computer model calibration using high-dimensional output. *Journal of the American Statistical Association* (2012)

[Höl03] HÖLLIG, Klaus: *Finite element methods with B-splines*, vol. 26, Siam (2003)

[Hom96] HOMMA, Toshimitsu and SALTELLI, Andrea: Importance measures in global sensitivity analysis of nonlinear models. *Reliability Engineering & System Safety* (1996), vol. 52(1):pp. 1–17

[How94] HOWELL, Peter D.: *Extensional thin layer flows*, Ph.D. thesis, University of Oxford (1994)

[Ioo15] IOOSS, Bertrand and LEMAÎTRE, Paul: A review on global sensitivity analysis methods, in: *Uncertainty Management in Simulation-Optimization of Complex Systems*, Springer (2015), pp. 101–122

[Ja11] JANYA-ANURAK, Chettapong; BERNARD, Thomas and BIRKENHOFER, Hannes: Numerical Sensitivity Analysis of a complex Glass Forming Process by means of local perturbations (2011), in: *Proceedings of COMSOL Conference, Stuttgart*

[Ja12] JANYA-ANURAK, Chettapong: Statistical inverse problem of partial differential equation: an example with stationary 1D heat conduction problem, Tech. Rep. IES-2011-15 (2012)

[Ja13a] JANYA-ANURAK, Chettapong: Inverse uncertainty quantification of a distributed parameter system: An application for glass forming model, Tech. Rep. IES-2012-10 (2013)

[Ja13b] JANYA-ANURAK, Chettapong; BIRKHOFER, Hannes; BERNARD, Thomas and QUILIN, Ma: 3D Multiphysics finite element model of a highly nonlinear glass forming process (2013), in: *NAFEMS World Congress 2013*

[Ja14] JANYA-ANURAK, Chettapong: Quantification of Uncertainties in a Distributed Parameter System using the Generalized Polynomial Chaos Expansion, Tech. Rep. IES-2013-03 (2014)

[Ja15a] JANYA-ANURAK, Chettapong; BERNARD, Thomas and BEYERER, Jürgen: A concept for sensitivity analysis and parameter calibration of coupled nonlinear PDEs and its application to an industrial glass forming model (2015), in: *1st ECCOMAS Thematic Conference on Uncertainty Quantification in Computational Sciences and Engineering (UNCECOMP 2015), Greece*, Elsevier, pp. 702–713

[Ja15b] JANYA-ANURAK, Chettapong; BIRKHOFER, Hannes; BERNARD, Thomas and QUILIN, Ma: 3D Multiphysik Finite Elemente Modell eines stark nichtlinearen Glasformungsprozesses. *NAFEMS Magazin, Online Journal, Zeitschrift für numerische Simulationsmethoden und angrenzende Gebiete* (2015), vol. 33:pp. 57–64

[Jay57] JAYNES, Edwin T.: Information theory and statistical mechanics. *Physical review* (1957), vol. 106(4):p. 620

[Jay68] JAYNES, Edwin T.: Prior probabilities. *IEEE Transactions on systems science and cybernetics* (1968), vol. 4(3):pp. 227–241

[Jay03] JAYNES, Edwin T.: *Probability theory: the logic of science*, Cambridge university press (2003)

[Jaz07] JAZWINSKI, Andrew H.: *Stochastic processes and filtering theory*, Courier Corporation (2007)

[Jul04] JULIER, Simon J. and UHLMANN, Jeffrey K.: Unscented filtering and nonlinear estimation. *Proceedings of the IEEE* (2004), vol. 92(3):pp. 401–422

[Kai05] KAIPIO, Jari and SOMERSALO, Erkki: *Statistical and computational inverse problems*, Springer, New York (2005)

[Kal60] KALMAN, Rudolph E.: A new approach to linear filtering and prediction problems. *Journal of basic Engineering* (1960), vol. 82(1):pp. 35–45

[Kap93] KAPUR, Jagat N.: *Maximum-entropy models in science and engineering*, Wiley, New York, revised ed. edn. (1993)

[Kee04] KEESE, Andreas: *Numerical Solutions of Systems with Stochastic Uncertainties: A General Purpose Framework for Stochastic Finite Elements*, Mechanik-Zentrum, Techn. Univ. (2004)

[Ken00] KENNEDY, Marc C. and O'HAGAN, Anthony: Bayesian Calibration of Computer Models. *Journal of the Royal Statistical Society, Series B, Methodological* (2000), vol. 63:pp. 425–464

[Kiu09] KIUREGHIAN, Armen Der and DITLEVSEN, Ove: Aleatory or epistemic? Does it matter? *Structural Safety* (2009), vol. 31(2):pp. 105–112

[Klo92] KLOEDEN, Peter E. and PLATEN, Eckhard: *Stochastic Differential Equations*, Springer Berlin Heidelberg, Berlin, Heidelberg (1992), pp. 103–160

[Ko95] KO, Jeonghwan; KURDILA, Andrew J. and PILANT, Michael S.: A class of finite element methods based on orthonormal, compactly supported wavelets. *Computational Mechanics* (1995), vol. 16(4):pp. 235–244

[Kro65] KRONROD, Aleksandr S.: *Nodes and weights of quadrature formulas: sixteen-place tables*, Not Avail (1965)

[Lee09] LEE, Sang H. and CHEN, Wei: A comparative study of uncertainty propagation methods for black-box-type problems. *Structural and Multidisciplinary Optimization* (2009), vol. 37(3):pp. 239–253

[Lew07] LEWANDOWSKI, Daniel; COOKE, Roger M. and TEBBENS, Radboud J. Duintjer: Sample-based estimation of correlation ratio with polynomial approximation. *ACM Transactions on Modeling and Computer Simulation (TOMACS)* (2007), vol. 18(1):p. 3

[Li11] LI, Han-Xiong and QI, Chenkun: *Spatio-temporal modeling of nonlinear distributed parameter systems: a time/space separation based approach*, vol. 50, Springer Science & Business Media (2011)

[Lie10] LIEBERMAN, Chad; WILLCOX, Karen and GHATTAS, Omar: Parameter and state model reduction for large-scale statistical inverse problems. *SIAM Journal on Scientific Computing* (2010), vol. 32(5):pp. 2523–2542

[LM10] LE MAÎTRE, Olivier and KNIO, Omar M.: *Spectral methods for uncertainty quantification: with applications to computational fluid dynamics*, Springer Science & Business Media (2010)

[Loc02] LOCH, Horst (Editor): *Mathematical simulation in glass technology,* Schott series on glass and glass ceramics, Springer, Berlin and Heidelberg (2002)

[Lue97] LUENBERGER, David G.: *Optimization by vector space methods,* John Wiley & Sons (1997)

[Mad06] MADSEN, Henrik O.; KRENK, Steen and LIND, Niels C.: *Methods of structural safety,* Courier Corporation (2006)

[Mah00] MAHADEVAN, Nagabhushan and HOO, Karlene A.: Wavelet-based model reduction of distributed parameter systems. *Chemical Engineering Science* (2000), vol. 55(19):pp. 4271–4290

[Mar09] MARZOUK, Youssef and XIU, Dongbin: A Stochastic Collocation Approach to Bayesian Inference in Inverse Problems. *Communications in Computational Physics* (2009), vol. 6:pp. 826–847

[Mat05] MATTHIES, Hermann G. and KEESE, Andreas: Galerkin methods for linear and nonlinear elliptic stochastic partial differential equations. *Computer Methods in Applied Mechanics and Engineering* (2005), vol. 194(12):pp. 1295–1331

[Mat07] MATTHIES, Hermann G.: QUANTIFYING UNCERTAINTY: MODERN COMPUTATIONAL REPRESENTATION OF PROBABILITY AND APPLICATIONS, in: Adnan Ibrahimbegovic and Ivica Kozar (Editors) *Extreme Man-Made and Natural Hazards in Dynamics of Structures,* NATO Security through Science Series, Springer Netherlands (2007), pp. 105–135

[Mat16a] MATTHIES, Hermann G.; ZANDER, Elmar; ROSIĆ, Bojana V. and LITVINENKO, Alexander: Parameter Estimation via Conditional Expectation—A Bayesian Inversion. *arXiv preprint arXiv:1606.09440* (2016)

[Mat16b] MATTHIES, Hermann G.; ZANDER, Elmar; ROSIĆ, Bojana V.; LITVINENKO, Alexander and PAJONK, Oliver: Inverse Problems in a Bayesian Setting, in: *Computational Methods for Solids and Fluids,* Springer (2016), pp. 245–286

[Maw08] MAWARDI, Andryas and PITCHUMANI, Ranga: Numerical simulations of an optical fiber drawing process under uncertainty. *Journal of Lightwave Technology* (2008), vol. 26(5):pp. 580–587

[McK00] McKay, Michael D.; Beckman, Richard J. and Conover, William J.: A comparison of three methods for selecting values of input variables in the analysis of output from a computer code. *Technometrics* (2000), vol. 42(1):pp. 55–61

[Mer09] Mercer, James: Functions of positive and negative type, and their connection with the theory of integral equations. *Philosophical transactions of the royal society of London. Series A, containing papers of a mathematical or physical character* (1909), vol. 209:pp. 415–446

[Mor92] Morgan, Millett G.; Henrion, Max and Small, Mitchell: *Uncertainty: a guide to dealing with uncertainty in quantitative risk and policy analysis*, Cambridge university press (1992)

[Mor95] Morokoff, William J. and Caflisch, Russel E.: Quasi-monte carlo integration. *Journal of computational physics* (1995), vol. 122(2):pp. 218–230

[Naj09] Najm, Habib N.: Uncertainty quantification and polynomial chaos techniques in computational fluid dynamics. *Annual Review of Fluid Mechanics* (2009), vol. 41:pp. 35–52

[Nei10] Neidinger, Richard D.: Introduction to automatic differentiation and MATLAB object-oriented programming. *SIAM review* (2010), vol. 52(3):pp. 545–563

[Nob01] Nobile, Fabio: Numerical approximation of fluid-structure interaction problems with application to haemodynamics (2001)

[Not97] Notaris, Sotirios E.: Interpolatory quadrature formulae with Chebyshev abscissae of the third or fourth kind. *Journal of computational and applied mathematics* (1997), vol. 81(1):pp. 83–99

[Nov96] Novak, Erich and Ritter, Klaus: High dimensional integration of smooth functions over cubes. *Numerische Mathematik* (1996), vol. 75(1):pp. 79–97

[Obe98a] Oberkampf, William L. and Blottner, Frederick G.: Issues in computational fluid dynamics code verification and validation. *AIAA Journal* (1998), vol. 36(5):pp. 687–695

[Obe98b] OBERKAMPF, William L.; SINDIR, Munir M. and CONLISK, A. Terrence: Guide for the verification and validation of computational fluid dynamics simulations. *Am. Institute of Aeronautics and Astronautics* (1998)

[Obe05] OBERKAMPF, William L. and HELTON, Jon C.: Evidence theory for engineering applications (2005)

[Obe10] OBERKAMPF, William L. and ROY, Christopher J.: *Verification and validation in scientific computing*, Cambridge university press (2010)

[Ola12] OLADYSHKIN, Sergey and NOWAK, Wolfgang: Data-driven uncertainty quantification using the arbitrary polynomial chaos expansion. *Reliability Engineering & System Safety* (2012), vol. 106:pp. 179–190

[Paj12] PAJONK, Oliver; ROSIĆ, Bojana V.; LITVINENKO, Alexander and MATTHIES, Hermann G.: A deterministic filter for non-Gaussian Bayesian estimation-applications to dynamical system estimation with noisy measurements. *Physica D: Nonlinear Phenomena* (2012), vol. 241(7):pp. 775–788

[Paj13] PAJONK, Oliver; ROSIĆ, Bojana V. and MATTHIES, Hermann G.: Sampling-free linear Bayesian updating of model state and parameters using a square root approach. *Computers & Geosciences* (2013), vol. 55:pp. 70–83

[Pat68] PATTERSON, T.N.L.: The optimum addition of points to quadrature formulae. *Mathematics of Computation* (1968), vol. 22(104):pp. 847–856

[Pes03] PESCH, Bernd: *Bestimmung der Messunsicherheit nach GUM*, Books on Demand, Norderstedt (2003)

[Puk93] PUKELSHEIM, Friedrich: *Optimal design of experiments*, vol. 50, siam (1993)

[Pye05] PYE, David; JOSEPH, Innocent and MONTENERO, Angelo: *Properties of glass-forming melts*, CRC Press (2005)

[Roa97] ROACHE, Patrick J.: Quantification of uncertainty in computational fluid dynamics. *Annual Review of Fluid Mechanics* (1997), vol. 29(1):pp. 123–160

[Ros24] ROSSELAND, Svein: Note on the absorption of radiation within a star. *Monthly Notices of the Royal Astronomical Society* (1924), vol. 84:pp. 525–528

[Ros12a] ROSIĆ, Bojana V.: *Variational formulations and functional approximation algorithms in stochastic plasticity of materials*, Ph.D. thesis, Faculty of Engineering , Kragujevac (2012)

[Ros12b] ROSIĆ, Bojana V.; LITVINENKO, Alexander; PAJONK, Oliver and MATTHIES, Hermann G.: Sampling-free linear Bayesian update of polynomial chaos representations. *Journal of Computational Physics* (2012), vol. 231(17):pp. 5761–5787

[Ros13] ROSIĆ, Bojana V.; KUČEROVÁ, Anna; SÝKORA, Jan; PAJONK, Oliver; LITVINENKO, Alexander and MATTHIES, Hermann G.: Parameter identification in a probabilistic setting. *Engineering Structures* (2013), vol. 50:pp. 179–196

[Roy11] ROY, Christopher J. and OBERKAMPF, William L.: A comprehensive framework for verification, validation, and uncertainty quantification in scientific computing. *Computer Methods in Applied Mechanics and Engineering* (2011), vol. 200(25):pp. 2131–2144

[Rus03] RUSSELL, Stuart Jonathan; NORVIG, Peter; CANNY, John F.; MALIK, Jitendra M. and EDWARDS, Douglas D.: *Artificial intelligence: a modern approach*, vol. 2, Prentice hall Upper Saddle River (2003)

[Sal04] SALTELLI, Andrea; TARANTOLA, Stefano; CAMPOLONGO, Francesca and RATTO, Marco: *Sensitivity analysis in practice: a guide to assessing scientific models*, John Wiley & Sons (2004)

[San13] SANTNER, Thomas J.; WILLIAMS, Brian J. and NOTZ, William I.: *The design and analysis of computer experiments*, Springer Science & Business Media (2013)

[Sär13] SÄRKKÄ, Simo: *Bayesian filtering and smoothing*, vol. 3, Cambridge University Press (2013)

[Saw09] SAWO, Felix and BERNARD, Thomas: Finite Element Model of a complex Glass Forming Process as a Tool for Control Optimization (2009), in: *European COMSOL Conference, Milan, Italy*

[Sch79] SCHLESINGER, Stewart; CROSBIE, Roy E; GAGNÉ, Roland E; INNIS, George S; LALWANI, Chandra S; LOCH, Joseph; SYLVESTER, Richard J; WRIGHT, Richard D; KHEIR, Naim and BARTOS, Dale: Terminology for model credibility. *Simulation* (1979), vol. 32(3):pp. 103–104

[Seo02] SEO, Hyun Seok and KWAK, Byung Man: Efficient statistical tolerance analysis for general distributions using three-point information. *International journal of production research* (2002), vol. 40(4):pp. 931–944

[Sha97] SHAMPINE, Lawrence F. and REICHELT, Mark W.: The matlab ode suite. *SIAM journal on scientific computing* (1997), vol. 18(1):pp. 1–22

[She05] SHELBY, James E.: *Introduction to glass science and technology*, Royal Society of Chemistry (2005)

[Sim89] SIMMONS, J.H. and SIMMONS, C.J.: Nonlinear viscous flow in glass forming. *American Ceramic Society Bulletin* (1989), vol. 68(11):pp. 1949–1955

[Smi13] SMITH, Ralph C.: *Uncertainty Quantification: Theory, Implementation, and Applications*, vol. 12, SIAM (2013)

[Smo63] SMOLYAK, Sergey A.: Quadrature and interpolation formulas for tensor products of certain classes of functions (1963), in: *Dokl. Akad. Nauk SSSR*, vol. 4, p. 123

[Sob90] SOBOL', Il'ya Meerovich: On sensitivity estimation for nonlinear mathematical models. *Matematicheskoe Modelirovanie* (1990), vol. 2(1):pp. 112–118, translated as "Sensitivity estimates for nonlinear mathematical models" in *Mathematical Modelling and Computational Experiments* (1993) 1(4): pp.407-414

[Soi08] SOIZE, Christian: Construction of probability distributions in high dimension using the maximum entropy principle: Applications to stochastic processes, random fields and random matrices. *International Journal for Numerical Methods in Engineering* (2008), vol. 76(10):pp. 1583–1611

[Sud00] SUDRET, Bruno and DER KIUREGHIAN, Armen: *Stochastic finite element methods and reliability: a state-of-the-art report*, Department of Civil and Environmental Engineering, University of California Berkeley, CA (2000)

[Sud08] SUDRET, Bruno: Global sensitivity analysis using polynomial chaos expansions. *Reliability Engineering & System Safety* (2008), vol. 93(7):pp. 964–979

[Tar05] TARANTOLA, Albert: *Inverse problem theory and methods for model parameter estimation*, siam (2005)

[Thr05] THRUN, Sebastian; BURGARD, Wolfram and FOX, Dieter: *Probabilistic robotics*, MIT press (2005)

[Tre08] TREFETHEN, Lloyd N.: Is Gauss quadrature better than Clenshaw-Curtis? *SIAM review* (2008), vol. 50(1):pp. 67–87

[Van83] VANMARCKE, Erik and GRIGORIU, Mircea: Stochastic finite element analysis of simple beams. *Journal of Engineering Mechanics* (1983), vol. 109(5):pp. 1203–1214

[vdL14] VON DER LINDEN, Wolfgang; DOSE, Volker and VON TOUSSAINT, Udo: *Bayesian probability theory: applications in the physical sciences*, Cambridge University Press (2014)

[vT11] VON TOUSSAINT, Udo: Bayesian inference in physics. *Reviews of Modern Physics* (2011), vol. 83(3):p. 943

[Wal03] WALKER, Warren E.; HARREMOËS, Poul; ROTMANS, Jan; VAN DER SLUIJS, Jeroen P.; VAN ASSELT, Marjolein B. A.; JANSSEN, Peter and KRAYER VON KRAUSS, Martin P.: Defining uncertainty: a conceptual basis for uncertainty management in model-based decision support. *Integrated assessment* (2003), vol. 4(1):pp. 5–17

[Wan97] WANG, J. and GADALA, Mohamed S.: Formulation and survey of ALE method in nonlinear solid mechanics. *Finite Elements in Analysis and Design* (1997), vol. 24(4):pp. 253–269

[Wan04] WANG, Jingbo and ZABARAS, Nicholas: A Bayesian inference approach to the inverse heat conduction problem. *International Journal of Heat and Mass Transfer* (2004), vol. 47(17):pp. 3927–3941

[Wie38] WIENER, Norbert: The Homogeneous Chaos. *American Journal of Mathematics* (1938), vol. 60(4):pp. 897–936

[Xiu02] XIU, Dongbin and KARNIADAKIS, George E.: The Wiener–Askey polynomial chaos for stochastic differential equations. *SIAM Journal on Scientific Computing* (2002), vol. 24(2):pp. 619–644

[Xiu04a] XIU, Dongbin: *Generalized (Wiener-Askey) Polynomial Chaos*, Ph.D. thesis, Brown University (2004)

[Xiu04b] XIU, Dongbin and KARNIADAKIS, George E.: Supersensitivity due to uncertain boundary conditions. *International journal for numerical methods in engineering* (2004), vol. 61(12):pp. 2114–2138

[Xiu05] XIU, Dongbin and HESTHAVEN, Jan S.: High-order collocation methods for differential equations with random inputs. *SIAM Journal on Scientific Computing* (2005), vol. 27(3):pp. 1118–1139

[Xiu10] XIU, Dongbin: *Numerical methods for stochastic computations: A spectral method approach*, Princeton University Press, Princeton and N.J (2010)

[Xu04] XU, Hequin and RAHMAN, Sharif: A generalized dimension-reduction method for multidimensional integration in stochastic mechanics. *International Journal for Numerical Methods in Engineering* (2004), vol. 61(12):pp. 1992–2019

Appendix

A Selected Fundamental Elements of Probability Theory and Stochastic Processes

This part of the appendix treats some selected fundamental aspects of probability, random field and statistics that are employed with in this thesis. The notation and the definition of the terms used are arranged in this appendix. The following statements are summarized from [LM10] and [Smi13]. Further elements of associated topics can be found in the references cited in the text.

A.1 Probability space and random variables

Definition 1 (Probability Space) A probability space is a three-tuple $(\Omega,\mathfrak{A},\mathcal{P})$ which consists of the components:

- Ω : **Sample Space**: is a nonempty set of all possible outcomes (from an experiment).
- $\mathfrak{A}$: **Event Space**: is a collection of possible events of interest. $\mathfrak{A}$ is also called the $\sigma-$algebra on Ω.

- $\mathcal{P}$: **Probability measure**: is a function $\mathcal{P} : \mathfrak{A} \rightarrow [0,1](\mathbb{R})$ that assigns probabilities to the events of $\mathfrak{A}$. The probability measure must satisfy the Kolmogorov axioms:
  1. Non-negativity $\forall \mathcal{A} \in \mathfrak{A} : \mathcal{P}(\mathcal{A}) \geq 0$
  2. Unitarity $\mathcal{P}(\Omega) = 1$
  3. $\sigma-$additivity $\forall \mathcal{A}_i, \mathcal{A}_j \in \mathfrak{A}$ and $\mathcal{A}_i \cap \mathcal{A}_j = \emptyset$ for $\forall\ i \neq j$, then

$$\mathcal{P}(\bigcup_{i=1}^{\infty} \mathcal{A}_i) = \sum_{i=1}^{\infty} \mathcal{P}(\mathcal{A}_i). \qquad \square$$

Definition 2 (Random Variable) Given a probability space $(\Omega, \mathfrak{A}, \mathcal{P})$, a random variable (RV) X is a mapping $X : (\Omega) \rightarrow \mathbb{R}$. □

Definition 3 (Cumulative Distribution Function) The cumulative distribution function (cdf) of a RV X is a mapping $\mathbb{R} \rightarrow [0,1]$ defined by

$$P_X(x) = \mathcal{P}\{\omega \in \Omega | X(\omega) \leq x\}. \qquad \square$$

Definition 4 (Probability Density Function) If P_X of a continuous RV X is absolutely continuous in $\mathbb{R}$, there is an integrable function p_X such that

$$P_X(x) = \int_{-\infty}^{x} p_X(\zeta) d\zeta,$$

where the derivative $p_X = \frac{dP_x}{dx}$ mapping $\mathbb{R}$ to $[0,\infty]$ is called the probability density function (pdf) of X. □

Definition 5 (Random Vector) The previous definition of a RV can be extended to $\mathbb{R}^d$-valued functions. Let $X_1, \ldots, X_d$ be random variables. The vector $\boldsymbol{X} : \Omega \rightarrow \mathbb{R}^d$ given by $\boldsymbol{X} = [X_1, \ldots, X_d]$ is called a random vector. □

Definition 6 (Joint Cumulative Distribution Function) For a random vector $\boldsymbol{X}$, the associated joint cdf $P_{\boldsymbol{X}} : \mathbb{R}^d \rightarrow [0,1]$ is defined by

$$P_{\boldsymbol{X}}(x_1, \ldots, x_d) = \mathcal{P}\left\{\omega \in \Omega | X_j(\omega) \leq x_j\right\}, j = 1, \ldots, d,$$

which is often written as $\mathcal{P}\{\omega \in \Omega | X_1 \leq x_1, \ldots, X_d \leq x_d\}$ □

Definition 7 (Marginal Probability Density Function) Let $X_1,\dots,X_d$ be jointly continuous random variables with joint pdf $p_{\mathbf{X}}(x_1,\dots,x_d)$. The marginal density functions of X_j is given by

$$p_{X_j}(x_j) = \int_{\mathbb{R}} \cdots \int_{\mathbb{R}} p_{\mathbf{X}}(x_1,\dots,x_d)dx_1 \cdots dx_{j-1}dx_{j+1} \cdots dx_d \qquad \square$$

Definition 8 (Conditional Probability Density Function) Let $X_1,\dots,X_d$ be jointly continuous RVs with joint pdf $p_{\mathbf{X}}(x_1,\dots,x_d)$ and the marginal density functions of X_j is $p_{X_j}(x_j)$. The conditional pdf of $X_1,\dots,X_{j-1},X_{j+1},\dots,X_d$ is

$$p_{X_1,\dots,X_{j-1},X_{j+1},\dots,X_d|X_j}(x_1,\dots,x_{j-1},x_{j+1},\dots,x_d|x_j) = \frac{p_{\mathbf{X}}(\boldsymbol{x})}{p_{X_j}(x_j)} \qquad \square$$

Definition 9 (Expectation) The expected value of the RV X is defined as the Lebesgue integral

$$\mathbb{E}[X] = \int_{\Omega} X d\mathcal{P} \qquad \square$$

Definition 10 (Moments of a Random Variable) Let X be a $\mathbb{R}$-valued random variable defined on the probability space $(\Omega,\mathfrak{A},\mathcal{P})$ with the CDF $P_X(x)$, and $Y = X^r$ for $r \in \mathbb{N}$. The expectation of Y is called the **moment of order** r of X, denoted $m_r(X)$:

$$m_r(X) = \mathbb{E}[X^r] = \int_{\Omega} x^r \, dP_X(x)$$

The $m_r(X)$ exists and is finite, if $X \in \mathcal{L}_r(\Omega,\mathfrak{A},\mathcal{P})$. If $Y = |X|^r$, the expectation of Y is called **absolute moment of order** r of X

$$\mathbb{E}[|X|^r] = \int_{\Omega} |X|^r \, dP_X(x)$$

The moment of order 1 of RV $m_r(X) = \mathbb{E}[X]$ is known as the mean $\mu_X = \overline{X}$ of the RV X. With the mean, the **central moment of order** r of X is defined as:

$$\mathbb{E}[X^r] = \int_{\Omega} (x - \mu_X)^r \, dP_X(x). \qquad \square$$

Definition 11 (Variance and Covariance) Given the random variable X, the variance of the RV X is defined as the central moment of order 2 of X

$$\mathrm{Var}[X] = \mathbb{E}\left[(X - \mathbb{E}(X))^2\right].$$

Let Y be another random variable, the covariance between the RV X and Y is defined as

$$\mathrm{Cov}[X,Y] = \mathbb{E}\left[(X - \mathbb{E}(X))(X - \mathbb{E}(Y))\right].$$

In the special case of $Y = X$

$$\mathrm{Cov}[X,X] = \mathbb{E}\left[(X - \mathbb{E}(X))(X - \mathbb{E}(X))\right] = \mathrm{Var}[X],$$

the covariance between the RV X and itself results the variance of the X. Considering the covariance between the random vector $\boldsymbol{X}$ and the random vector $\boldsymbol{Y}$, relation (11) is generalized into the **cross-covariance matrix**

$$\mathrm{Cov}[\boldsymbol{X},\boldsymbol{Y}] = \mathbf{C}_{\boldsymbol{XY}} = \mathbb{E}\left[(\boldsymbol{X} - \mathbb{E}(\boldsymbol{X}))(\boldsymbol{Y} - \mathbb{E}(\boldsymbol{Y}))^T\right], \tag{A.1}$$

and the **(auto)-covariance matrix** in the special case of $\boldsymbol{Y} = \boldsymbol{X}$

$$\mathrm{Cov}[\boldsymbol{X},\boldsymbol{X}] = \mathbf{C}_{\boldsymbol{XX}} = \mathbb{E}\left[(\boldsymbol{X} - \mathbb{E}(\boldsymbol{X}))(\boldsymbol{X} - \mathbb{E}(\boldsymbol{X}))^T\right]. \tag{A.2}$$

The variance of the random vector $\boldsymbol{X}$ is defined as

$$\mathrm{Var}[\boldsymbol{X}] = \mathbb{E}\left[(\boldsymbol{X} - \mathbb{E}(\boldsymbol{X}))^2\right]\mathbb{E}\left[(\boldsymbol{X} - \mathbb{E}(\boldsymbol{X}))^T(\boldsymbol{X} - \mathbb{E}(\boldsymbol{X}))\right].$$

Indeed, the entire diagonal elements of the covariance matrix $\mathbf{C}_{\boldsymbol{XX}}$ are the variance of each element of the random vector $\boldsymbol{X}$. $\square$

For the discussion of the covariance with the PCE, equations (A.1) and (A.2) can be defined with the tensor product:

$$\mathbf{C}_{XY} = \mathbb{E}\left[\tilde{\boldsymbol{X}} \otimes \tilde{\boldsymbol{Y}}\right],$$
$$\mathbf{C}_{XX} = \mathbb{E}\left[\tilde{\boldsymbol{X}} \otimes \tilde{\boldsymbol{X}}\right],$$

where $\tilde{\boldsymbol{X}} = \boldsymbol{X} - \mathbb{E}(\boldsymbol{X})$ and $\tilde{\boldsymbol{Y}} = \boldsymbol{Y} - \mathbb{E}(\boldsymbol{Y})$ are defined as the fluctuations of the random vector $\boldsymbol{X}$ and $\boldsymbol{Y}$. The tensor product $\boldsymbol{a} \otimes \boldsymbol{b}$ is equivalent to a matrix multiplication $\boldsymbol{ab}^T$, e.g. for $m \times 1$ column vector a and $n \times 1$ column vector b, it yields

$$\boldsymbol{a} \otimes \boldsymbol{b} = \boldsymbol{ab}^T = \begin{bmatrix} a_1 \\ \vdots \\ a_i \\ \vdots \\ a_m \end{bmatrix} \begin{bmatrix} b_1 & \cdots & b_j & \cdots & b_n \end{bmatrix} = \begin{bmatrix} a_1 b_1 & \cdots & a_1 b_n \\ \vdots & \ddots & \vdots \\ a_m b_1 & \cdots & a_m b_n \end{bmatrix}.$$

A.2 Convergence of random variables

Let X be a $\mathbb{R}$-valued RV and $X_{n \geq 1}$ be a sequence of RVs defined on a probability space $(\Omega, \mathfrak{A}, \mathcal{P})$. The convergence of the sequence X_n to X depends on the way $X - X_n$ is measured. Alternatives of the convergences are:

- Almost sure convergence,

$$X_n \xrightarrow{\text{a.s.}} X \qquad \text{if } \lim_{n\to\infty} X_n(\omega) = X(\omega), \forall \omega \in \Omega \backslash N, \mathcal{P}(N) = 0.$$

- Convergence in **probability**,

$$X_n \xrightarrow{\text{pr}} X \qquad \text{if } \lim_{n\to\infty} \mathcal{P}\left(|X_n(\omega) - X(\omega)| > \varepsilon\right) = 0, \forall \varepsilon > 0.$$

- Convergence in **distribution**,

$$X_n \xrightarrow{\text{dis}} X \qquad \text{if } \lim_{n\to\infty} P_{X_n}(x) = P_X(x), \forall x \in \mathbb{R}.$$

- Convergence in $\mathcal{L}_p$

$$X_n \xrightarrow{\text{m.p.}} X \qquad \text{if } \lim_{n\to\infty} \mathbb{E}\left[|X_n - X|^p\right] = 0.$$

- The mean square convergence is the convergence in $\mathcal{L}_2$

$$X_n \xrightarrow{\text{m.s.}} X \qquad \text{if } \lim_{n\to\infty} \mathbb{E}\left[|X_n - X|^2\right] = 0.$$

A.3 Stochastic processes and random fields

The stochastic process and the random field can be seen as extensions of the notion of random variable and random vector that incorporate a dependence on time or/and space coordinates. The random vector is defined as $\mathbb{R}^d$-valued random variable, which is the mapping $\boldsymbol{X} : (\Omega) \to \mathbb{R}^d$. In the context of functional analysis, the $\mathbb{R}^d$-valued RV can be generalized to the $\mathcal{V}-$**valued random variable** with the following definition.

Definition 12 ($\mathcal{V}$-valued Random Variable) Given a probability space $(\Omega,\mathfrak{A},\mathcal{P})$ and a vector space $\mathcal{V}$, a $\mathcal{V}$-valued random variable $X_{\mathcal{V}}$ is a measurable function on probability space $X_{\mathcal{V}} : (\Omega) \to \mathcal{V}$, such that $\{\omega \in \Omega | X(\omega) \leq x\} \in \mathfrak{A}$ for each $x \in \mathcal{V}$. □

For later discussion, it will be useful to view the space of all $\mathcal{V}$-valued RVs as linear combinations of elements $X(\omega)\boldsymbol{v}$, where $X(\omega)$ is a real-valued RV, and $\boldsymbol{v} \in \mathcal{V}$. In order words, the $\mathcal{V}$-valued RVs are elements of tensor product $\mathcal{L}_2(\Omega) \otimes \mathcal{V}$, where $\mathcal{L}_2(\Omega)$ is the space of real-valued random variables. With the definition of the $\mathcal{V}$-valued random variable, the **stochastic process** can be defined as:

Definition 13 (Stochastic Process) Given a probability space $(\Omega,\mathfrak{A},\mathcal{P})$,
A stochastic process is a collection of $\mathcal{V}-$valued random variables on Ω, indexed by a totally ordered set $\mathcal{T}$. That is, a stochastic process $\mathcal{X}$ is a collection

$$\{\mathcal{X}(t;\cdot) : t \in \mathcal{T}\}$$

where each $\mathcal{X}(t;\cdot)$ is an $\mathcal{V}-$valued random variable on Ω. □

The set $\mathcal{T}$ is some interval, but mostly it is referred to the time interval $\mathcal{T} = [0, t_{\text{end}}]$, because of the application viewpoint. A stochastic process can be interpreted as follows:

1. $\mathcal{X}(t;\omega)$ is a function on $\mathcal{T} \times \Omega$ with the realization $x(t)$ for given $t \in \mathcal{T}$ and $\omega \in \Omega$.
2. For a fixed time $t \in \mathcal{T}$, $\mathcal{X}(t;\cdot)$ is determined as a random variable.
3. For an outcome $\omega \in \Omega$, the realization of $\mathcal{X}(\cdot;\omega) = x(t)$ is a function of t, which is called a *trajectory*, a *sample path* associated with ω.

Another representation of the stochastic process is proposed by Matthies [Mat07]. Considering the vector space $\mathcal{V}$ as a space of of values of function $\mathcal{F}(\mathcal{T})$ on the interval $\mathcal{T}$, the $\mathcal{F}(\mathcal{T})$-valued random variable $X_{\mathcal{F}(\mathcal{T})}$ is a function $\mathcal{X} : \mathcal{T} \times \Omega \to \mathcal{F}(\mathcal{T})$ with argument -time- $t \in \mathcal{T}$ and -event- $\omega \in \Omega$. The function $\mathcal{X}(t;\omega) \in \mathcal{L}_2(\omega) \otimes \mathcal{F}(\mathcal{T})$ is then called a **stochastic process.**

Instead of one-dimensional interval $\mathcal{T}$, the multidimensional domain like spatial domain $\mathcal{G}$ can be also considered. In case that the space $\mathcal{V}$ is a space of function $\mathcal{F}(\mathcal{G})$, the function $\mathcal{X} : \mathcal{G} \times \Omega \to \mathcal{F}(\mathcal{G})$ is termed as a **random field**. The description of random fields $\mathcal{X}(\boldsymbol{r},\omega)$, where the random variable is assigned to each point $\boldsymbol{r} \in \mathcal{G}$ in the spatial domain $\mathcal{G}$, is similar to the context of stochastic process. The concept of stochastic processes should be generalized so that the underlying argument can be multidimensional vectors or points on a manifold.

Definition 14 (Random Field) Given a probability space $(\Omega,\mathfrak{A},\mathcal{P})$,
An $\mathcal{F}(\mathcal{G})-$valued random field is a collection of $\mathcal{F}(\mathcal{G})-$valued random variables on Ω, indexed by elements in a topological space $\mathcal{G}$. That is, a random field $\mathcal{X}$ is a collection

$$\{\mathcal{X}(\boldsymbol{r};\cdot) : \boldsymbol{r} \in \mathcal{G}\},$$

where each $\mathcal{X}(\boldsymbol{r};\cdot)$ is an $\mathcal{F}(\mathcal{G})-$valued random variable. □

Definition 15 (Covariance Function) The covariance function, or sometimes known as kernel, describes the spatial covariance of a random field. The definition of the covariance (Def.11) can be generalized for the random field by means of the tensor product. The (auto)-covariance function of the random field $\mathcal{X}(\boldsymbol{r},\omega)$ is given by:

$$\mathrm{Cov}_{\mathcal{X}}(\boldsymbol{r},\boldsymbol{r}') := \mathbb{E}\left[\tilde{\mathcal{X}}(\boldsymbol{r},\omega) \otimes \tilde{\mathcal{X}}(\boldsymbol{r}',\omega)\right], \tag{A.3}$$

□

where $\tilde{\mathcal{X}}(\boldsymbol{r},\omega) = \mathcal{X}(\boldsymbol{r},\omega) - \overline{\mathcal{X}}(\boldsymbol{r})$ is the fluctuation part of the random field.

B Probability Distributions in Generalized Polynomial Chaos

B.1 Uniform distribution

- **Probability density function**

$$p_\xi(\xi) = \begin{cases} \frac{1}{b-a}, & \text{for } a < \xi < b, \\ 0, & \text{otherwise} \end{cases}$$

- **Associated quadrature rule**: Gauss-Lengendre, Clenshaw-Curtis or Gauss-Patterson
  The integration of a function $f(\zeta)$ has the standard form

$$I = \int_{-1}^{1} f(\zeta)d\zeta$$

- **Transformation function and its derivative**

$$\xi = \chi(\zeta) = \left(\frac{b-a}{2}\right)\zeta + \left(\frac{b+a}{2}\right)$$
$$\left|\frac{d\xi}{d\zeta}\right| = \left(\frac{b-a}{2}\right)$$

- **Integration with the transformed function**

$$I = \int_{\xi=a}^{\xi=b} f(\xi)d\xi = \left(\frac{b-a}{2}\right) \int_{\zeta=-1}^{\zeta=1} f(\chi(\zeta))p_\xi(\chi(\zeta))d\zeta$$

B.2 Normal distribution

- **Probability density function**

$$p_\xi(\xi) = \frac{1}{\sigma\sqrt{2\pi}} \exp\left(-\frac{(\xi-\mu)^2}{2\sigma^2}\right)$$

- **Associated quadrature rule**: Gauss-Hermite
  The integration of a function $f(\zeta)$ has the standard form

$$I = \int_{-\infty}^{\infty} f(\zeta) e^{-\zeta^2} d\xi$$

- **Transformation function and its derivative**

$$\xi = \chi(\zeta) = \sqrt{2}\sigma\zeta + \mu$$
$$\left|\frac{d\xi}{d\zeta}\right| = \sqrt{2}\sigma$$

- **Integration with the transformed function**

$$I = \int_{\xi=-\infty}^{\xi=\infty} f(\xi) p_\xi(\xi) d\xi = \frac{1}{\sqrt{\pi}} \int_{\zeta=-\infty}^{\zeta=\infty} e^{-\zeta^2} f(\chi(\zeta)) d\zeta$$

B.3 Exponential distribution

- **Probability density function**

$$p_\xi(\xi) = \frac{1}{b} \exp(\frac{-\xi}{b})$$

- **Associated quadrature rule**: Gauss-Laguerre
  The integration of a function $f(\zeta)$ has the standard form

$$I = \int_{0}^{\infty} \exp(-\zeta) f(\zeta) d\zeta$$

- **Transformation function and its derivative**

$$\xi = \chi(\zeta) = b \cdot \zeta$$
$$\left|\frac{d\xi}{d\zeta}\right| = b$$

- **Integration with the transformed function**

$$I = \int_{\xi=0}^{\xi=\infty} f(\xi) p_\xi(\xi) d\xi = \int_{\zeta=0}^{\zeta=\infty} f(\chi(\zeta)) \exp(-\zeta) d\zeta$$

B.4 Gamma distribution

- **Probability density function**

$$p_\xi(\xi) = \frac{1}{\Gamma(a) b^a} \xi^{a-1} \exp(\frac{-\xi}{b})$$

where $\Gamma(a)$ is a Gamma function defined as

$$\Gamma(a) = \int_0^\infty x^{a-1} e^{-x} dx$$

in case that a is positive integer, it yields:

$$\Gamma(a) = (a-1)!$$

- **Associated quadrature rule**: Generalized Gauss-Laguerre
  The integration of a function $f(\zeta)$ has the standard form

$$I = \int_0^\infty \zeta^c \exp(-\zeta) f(\zeta) d\zeta$$

- **Transformation function and its derivative**

$$\xi = \chi(\zeta) = b \cdot \zeta$$

$$\left| \frac{d\xi}{d\zeta} \right| = b$$

- **Integration with the transformed function**

$$I = \int_{\xi=0}^{\xi=\infty} f(\xi) p_\xi(\xi) d\xi = \frac{1}{\Gamma(c+1)} \int_{\zeta=0}^{\zeta=\infty} f(\chi(\zeta))\zeta^c \exp(-\zeta) d\zeta$$

The exponential distribution can be considered as a special case of Gamma distribution, where $a = 1$.

B.5 Beta distribution

- **Probability density function**

$$p_\xi(\xi) = \frac{(\xi - a)^{q-1} (b - \xi)^{r-1}}{B(q,r)(b-a)}$$

where $B(q,r)$ is a Beta function defined as

$$B(q,r) = \frac{\Gamma(q)\Gamma(r)}{\Gamma(q+r)}$$

in case that q and r are positive integers, it yields

$$B(q,r) = \frac{(q-1)!(r-1)!}{(q+r-1)!}$$

- **Associated quadrature rule**: Gauss-Jacobi
  The integration of a function $f(\zeta)$ has the standard form

$$I = \int_{-1}^{1} f(\zeta)(1-\zeta)^\alpha (1+\zeta)^\beta d\zeta$$

- **Transformation function and its derivative**

$$\xi = \chi(\zeta) = \left(\frac{b-a}{2}\right)\zeta + \left(\frac{b+a}{2}\right)$$

$$\left|\frac{d\xi}{d\zeta}\right| = \left(\frac{b-a}{2}\right)$$

- **Integration with the transformed function**

$$I = \int\limits_{\xi=a}^{\xi=b} f(\xi) p_{\xi}(\xi) d\xi$$

$$= \frac{1}{B(q,r)} \int\limits_{\zeta=-1}^{\zeta=1} f(\chi(\zeta))\,(\zeta - a)^{q-1}\,(b-\zeta)^{r-1}\,d\zeta$$

The Beta distribution with the parameter $q = r = 1$ is equivalent to the uniform distribution.

C Orthogonal Polynomials

This part of the appendix treats some selected fundamentals about the orthogonal polynomial used in generalized polynomial chaos expansion. The following statements are summarized from [LM10] and [Xiu10]. Only the orthogonal polynomials applied in this thesis are summarized in this appendix. Further elements of associated topics can be found in the references cited in the text.

Considering continuous polynomials defined on the interval $a \leq \xi \leq b$. The square of the $\mathcal{L}_2$ norm of Ψ_k denoted by γ_k is given by the inner product with the weight $w(\xi)$ as

$$\gamma_k = \|\Psi_k\|^2 = \langle \Psi_k | \Psi_k \rangle \equiv \int \Psi_k^2(\xi) w(\xi) d\xi. \tag{C.1}$$

As discussed in [LM10] and [Xiu10], the orthogonal polynomials:

- satisfy the differential equation

  $$g_2(\xi)\Psi_k'' + g_1(\xi)\Psi_k' + c_k \Psi_k = 0 \tag{C.2}$$

  where g_1 and g_2 are independent of k and the c_k are constants that depend on k only.

- can be generated using **Rodrigues'** formula

  $$\Psi_k(\xi) = \frac{1}{e_k w(\xi)} \frac{d^k}{d\xi^k} \left[w(\xi)(g(\xi)^k) \right] \tag{C.3}$$

  where g is a polynomial in ξ that is independent of k and the e_k are arbitrary normalization factors that depend on k only.

- satisfy a three-term recurrence relation

  $$\Psi_{k+1} = (A_k \xi + B_k)\Psi_k - C_k \Psi_{k-1} \tag{C.4}$$

C.1 Legendre polynomials $Le_k(\xi)$

- Orthogonality
  The Legendre polynomials $Le_k(\xi)$ form an orthonormal relation

$$\langle Le_k(\xi)|Le_l(\xi)\rangle = \int Le_k(\xi)Le_l(\xi)\,w(\xi)\,d\xi = \delta_{kl}\gamma_k, \tag{C.5}$$

where $w(\xi) = 1/2$, with

$$\gamma_k = \int_{-1}^{1} Le_k^2(\xi)\,w(\xi)\,d\xi = \frac{1}{2k+1}, \tag{C.6}$$

- Differential Equation:

$$g_2(\xi) = 1-\xi^2, \qquad g_1(\xi) = -2\xi \qquad \text{and } c_k = k(k+1) \tag{C.7}$$

- Rodriguez formula

$$g(\xi) = 1-\xi^2, \qquad \text{and } e_k = (-1)^k 2^k k! \tag{C.8}$$

- Recurrence relation

$$Le_{k+1} = \frac{2k+1}{k+1}\xi Le_k - \frac{k}{k+1}Le_{k-1} \tag{C.9}$$

- First seven Legendre polynomials are given by

$$
\begin{aligned}
Le_0(\xi) &= 1,\\
Le_1(\xi) &= \xi,\\
Le_2(\xi) &= \frac{1}{2}\left(3\xi^2-1\right),\\
Le_3(\xi) &= \frac{1}{2}\left(5\xi^3-3\xi\right),\\
Le_4(\xi) &= \frac{1}{8}\left(35\xi^4-30\xi^2+3\right),\\
Le_5(\xi) &= \frac{1}{8}\left(63\xi^5-70\xi^2+15\xi\right),\\
Le_6(\xi) &= \frac{1}{16}\left(231\xi^6-315\xi^4+105\xi^2-5\right).
\end{aligned}
$$

C.2 Hermite polynomials $H_k(\xi)$

- Orthogonality
  The Hermite polynomials $H_k(\xi)$ form an orthonormal relation

$$\langle H_k(\xi)|H_l(\xi)\rangle = \int H_k(\xi)H_l(\xi)w(\xi)d\xi = \delta_{kl}\gamma_k, \tag{C.10}$$

where $w(\xi) = \frac{1}{\sqrt{2\pi}}\exp\frac{-\xi^2}{2}$, with

$$\gamma_k = \int_{-1}^{1} H_k^2(\xi)w(\xi)d\xi = k!, \tag{C.11}$$

- Differential Equation

$$g_2(\xi) = 1, \qquad g_1(\xi) = -\xi \qquad \text{and } c_k = k \tag{C.12}$$

- Rodriguez formula

$$g(\xi) = 1, \qquad \text{and } e_k = (-1)^k. \tag{C.13}$$

- Recurrence relation

$$H_{k+1} = \xi H_k - kH_{k-1} \tag{C.14}$$

- First seven Hermite polynomials are given by

$$\begin{aligned}
H_0(\xi) &= 1,\\
H_1(\xi) &= \xi,\\
H_2(\xi) &= \xi^2 - 1,\\
H_3(\xi) &= \xi^3 - 3\xi,\\
H_4(\xi) &= \xi^4 - 6\xi^2 + 3,\\
H_5(\xi) &= \xi^5 - 10\xi^3 + 15\xi,\\
H_6(\xi) &= \xi^6 - 15\xi^4 + 45\xi^2 - 15.
\end{aligned}$$

C.3 Laguerre polynomials $La_k(\xi)$

- Orthogonality
  The Laguerre polynomials $La_k(\xi)$ form an orthonormal relation

$$\langle La_k(\xi)|La_l(\xi)\rangle = \int La_k(\xi)La_l(\xi)\,w(\xi)\,d\xi = \delta_{kl}\gamma_k, \tag{C.15}$$

where $w(\xi) = \exp(-\xi)$, with

$$\gamma_k = \int_{-1}^{1} La_k^2(\xi)\,w(\xi)\,d\xi = \frac{1}{2k+1}, \tag{C.16}$$

- Differential Equation

$$g_2(\xi) = \xi, \qquad g_1(\xi) = 1-\xi \qquad \text{and } c_k = k \tag{C.17}$$

- Rodriguez formula

$$g(\xi) = \xi, \qquad \text{and } e_k = k!. \tag{C.18}$$

- Recurrence relation

$$La_{k+1} = \frac{2k+1-\xi}{k+1}La_k - \frac{k}{k+1}La_{k-1} \tag{C.19}$$

- First seven Laguerre polynomials are given by

$$\begin{aligned}
La_0(\xi) &= 1,\\
La_1(\xi) &= -\xi+1,\\
La_2(\xi) &= \frac{1}{2}\left(\xi^2-4\xi+2\right),\\
La_3(\xi) &= \frac{1}{6}\left(-\xi^3+9\xi^2-18\xi+6\right),\\
La_4(\xi) &= \frac{1}{24}\left(\xi^4-16\xi^3+72\xi^2-96\xi+24\right),\\
La_5(\xi) &= \frac{1}{120}\left(-\xi^5+25\xi^4-200\xi^3+600\xi^2-600\xi+120\right),\\
La_6(\xi) &= \frac{1}{720}\left(\xi^6-36\xi^5+450\xi^4-2400\xi^3+5400\xi^2-4320\xi+720\right).
\end{aligned}$$

Karlsruher Schriftenreihe zur Anthropomatik (ISSN 1863-6489)

Herausgeber: Prof. Dr.-Ing. Jürgen Beyerer

Die Bände sind unter www.ksp.kit.edu als PDF frei verfügbar oder als Druckausgabe bestellbar.

Band 1 Jürgen Geisler
Leistung des Menschen am Bildschirmarbeitsplatz. 2006
ISBN 3-86644-070-7

Band 2 Elisabeth Peinsipp-Byma
Leistungserhöhung durch Assistenz in interaktiven Systemen zur Szenenanalyse. 2007
ISBN 978-3-86644-149-1

Band 3 Jürgen Geisler, Jürgen Beyerer (Hrsg.)
Mensch-Maschine-Systeme. 2010
ISBN 978-3-86644-457-7

Band 4 Jürgen Beyerer, Marco Huber (Hrsg.)
Proceedings of the 2009 Joint Workshop of Fraunhofer IOSB and Institute for Anthropomatics, Vision and Fusion Laboratory. 2010
ISBN 978-3-86644-469-0

Band 5 Thomas Usländer
Service-oriented design of environmental information systems. 2010
ISBN 978-3-86644-499-7

Band 6 Giulio Milighetti
Multisensorielle diskret-kontinuierliche Überwachung und Regelung humanoider Roboter. 2010
ISBN 978-3-86644-568-0

Band 7 Jürgen Beyerer, Marco Huber (Hrsg.)
Proceedings of the 2010 Joint Workshop of Fraunhofer IOSB and Institute for Anthropomatics, Vision and Fusion Laboratory. 2011
ISBN 978-3-86644-609-0

Band 8 Eduardo Monari
Dynamische Sensorselektion zur auftragsorientierten Objektverfolgung in Kameranetzwerken. 2011
ISBN 978-3-86644-729-5

Band 9 Thomas Bader
Multimodale Interaktion in Multi-Display-Umgebungen. 2011
ISBN 3-86644-760-8

Band 10 Christian Frese
Planung kooperativer Fahrmanöver für kognitive Automobile. 2012
ISBN 978-3-86644-798-1

Band 11 Jürgen Beyerer, Alexey Pak (Hrsg.)
Proceedings of the 2011 Joint Workshop of Fraunhofer IOSB and Institute for Anthropomatics, Vision and Fusion Laboratory. 2012
ISBN 978-3-86644-855-1

Band 12 Miriam Schleipen
Adaptivität und Interoperabilität von Manufacturing Execution Systemen (MES). 2013
ISBN 978-3-86644-955-8

Band 13 Jürgen Beyerer, Alexey Pak (Hrsg.)
Proceedings of the 2012 Joint Workshop of Fraunhofer IOSB and Institute for Anthropomatics, Vision and Fusion Laboratory. 2013
ISBN 978-3-86644-988-6

Band 14 Hauke-Hendrik Vagts
Privatheit und Datenschutz in der intelligenten Überwachung: Ein datenschutzgewährendes System, entworfen nach dem „Privacy by Design" Prinzip. 2013
ISBN 978-3-7315-0041-4

Band 15 Christian Kühnert
Data-driven Methods for Fault Localization in Process Technology. 2013
ISBN 978-3-7315-0098-8

Band 16 Alexander Bauer
Probabilistische Szenenmodelle für die Luftbildauswertung. 2014
ISBN 978-3-7315-0167-1

Band 17 Jürgen Beyerer, Alexey Pak (Hrsg.)
Proceedings of the 2013 Joint Workshop of Fraunhofer IOSB and Institute for Anthropomatics, Vision and Fusion Laboratory. 2014
ISBN 978-3-7315-0212-8

Band 18 Michael Teutsch
Moving Object Detection and Segmentation for Remote Aerial Video Surveillance. 2015
ISBN 978-3-7315-0320-0

Band 19 Marco Huber
Nonlinear Gaussian Filtering: Theory, Algorithms, and Applications. 2015
ISBN 978-3-7315-0338-5

Band 20 Jürgen Beyerer, Alexey Pak (Hrsg.)
Proceedings of the 2014 Joint Workshop of Fraunhofer IOSB and Institute for Anthropomatics, Vision and Fusion Laboratory. 2014
ISBN 978-3-7315-0401-6

Band 21 Todor Dimitrov
Permanente Optimierung dynamischer Probleme der Fertigungssteuerung unter Einbeziehung von Benutzerinteraktionen. 2015
ISBN 978-3-7315-0426-9

Band 22 Benjamin Kühn
Interessengetriebene audiovisuelle Szenenexploration. 2016
ISBN 978-3-7315-0457-3

Band 23 Yvonne Fischer
Wissensbasierte probabilistische Modellierung für die Situationsanalyse am Beispiel der maritimen Überwachung. 2016
ISBN 978-3-7315-0460-3

Band 24 Jürgen Beyerer, Alexey Pak (Hrsg.)
Proceedings of the 2015 Joint Workshop of Fraunhofer IOSB and Institute for Anthropomatics, Vision and Fusion Laboratory. 2016
ISBN 978-3-7315-0519-8

Band 25 Pascal Birnstill
Privacy-Respecting Smart Video Surveillance Based on Usage Control Enforcement. 2016
ISBN 978-3-7315-0538-9

Band 26 Philipp Woock
Umgebungskartenschätzung aus Sidescan-Sonardaten für ein autonomes Unterwasserfahrzeug. 2016
ISBN 978-3-7315-0541-9

Band 27 Janko Petereit
Adaptive State × Time Lattices: A Contribution to Mobile Robot Motion Planning in Unstructured Dynamic Environments. 2017
ISBN 978-3-7315-0580-8

Band 28 Erik Ludwig Krempel
Steigerung der Akzeptanz von intelligenter Videoüberwachung in öffentlichen Räumen. 2017
ISBN 978-3-7315-0598-3

Band 29 Jürgen Moßgraber
Ein Rahmenwerk für die Architektur von Frühwarnsystemen. 2017
ISBN 978-3-7315-0638-6

Band 30 Andrey Belkin
World Modeling for Intelligent Autonomous Systems. 2017
ISBN 978-3-7315-0641-6

Band 31 Chettapong Janya-Anurak
Framework for Analysis and Identification of Nonlinear Distributed Parameter Systems using Bayesian Uncertainty Quantification based on Generalized Polynomial Chaos. 2017
ISBN 978-3-7315-0642-3